HOW **NOT** TO STUDY
A DISEASE

HOW **NOT** TO STUDY A DISEASE

THE STORY OF ALZHEIMER'S

KARL HERRUP

THE MIT PRESS   CAMBRIDGE, MASSACHUSETTS   LONDON, ENGLAND

This book was set in Stone Serif and Avenir by Westchester Publishing Services. Printed and bound in the United States of America.

Library of Congress Cataloging-in-Publication Data

Names: Herrup, Karl, author.
Title: How not to study a disease : the story of Alzheimer's / Karl Herrup.
Description: Cambridge, Massachusetts : The MIT Press, [2021] |
    Includes bibliographical references and index.
Identifiers: LCCN 2020040776 | ISBN 9780262045902 (hardcover)
Subjects: MESH: Alzheimer Disease | Models, Biological
Classification: LCC RC523 | NLM WT 155 | DDC 616.8/311--dc23
LC record available at https://lccn.loc.gov/2020040776

10 9 8 7 6 5 4 3 2 1

This book is dedicated to Dorothy and her daughter and all of the millions of others whose lives have been touched by Alzheimer's disease. It is you who bring me into work every day.

I dedicate this book as well to my teachers from whom I have learned so much and to all of my students who have taught me so much.

CONTENTS

PROLOGUE   ix

I   IN THE BEGINNING   1

   1   A LAYPERSON'S HISTORY OF ALZHEIMER'S DISEASE   3
   2   A PHYSICIAN'S HISTORY OF ALZHEIMER'S DISEASE   23
   3   A SCIENTIST'S HISTORY OF ALZHEIMER'S DISEASE   33
   4   MYSTERY SOLVED! HOW FOUR DISCOVERIES TRANSFORMED AN ENTIRE FIELD   53

II   WHAT HAPPENED TO OUR CURE?   71

   5   BUILDING A MODEL OF ALZHEIMER'S DISEASE   73

III   DOUBLE-EDGED SWORDS   97

   6   FEDERAL SUPPORT OF BASIC BIOMEDICAL RESEARCH   99
   7   THE PHARMACEUTICAL AND BIOTECH INDUSTRY   115
   8   TESTING OUR MODELS: BREAKING BAD   131
   9   WHAT IS ALZHEIMER'S DISEASE?   149

IV   WHERE SHALL WE GO FROM HERE?   169

   10   A LAYPERSON'S GUIDE TO THE BIOLOGY OF AGING   171
   11   BUILDING A NEW MODEL OF ALZHEIMER'S DISEASE   187
   12   REBALANCING OUR RESEARCH PORTFOLIO   207

13 REBALANCING OUR INSTITUTIONS 227
14 FINAL THOUGHTS 243

ACKNOWLEDGMENTS 247
NOTES 249
INDEX 255

PROLOGUE

One of the difficult parts of writing this book is that I am really writing it for two different audiences. The first is a small audience of professional people currently working in the Alzheimer's disease field. These are my colleagues—physicians, scientists, members of lay organizations, and the press. Most of the story will be familiar to them, but some parts will be new and others may have uncomfortable implications for their own work in the field. For this audience, I have put in just a bit more scientific detail than I might have otherwise.

Yet while I care what my colleagues think about my ideas, the most important audience I am trying to reach is not made up of scientists. This second, larger group is made up of people in all walks of life that have been touched in some way by Alzheimer's disease. I have spoken to many such people over the years—family, friends, and total strangers. This book is for them. Because of that, I have done my best to keep the mind-numbing minutiae of the science (that I personally love) to a minimum. If you belong to this second group, I urge you not to get bogged down or overwhelmed by the details. I find each of the topics fascinating, and I hope that you find reading about each of them to be fun and educational. But the important message of the book is not about the science itself. It's about the people, and the good and bad decisions they made trying to find a cure for Alzheimer's. Of course, the professor in me hopes to teach as well as to entertain, but if you find yourself getting frustrated, skip ahead; you have my permission. My goal is to take you on a fascinating journey through my world—the world of Alzheimer's disease research—and help you understand where we are in our struggle to find a cure.

I

IN THE BEGINNING

1

A LAYPERSON'S HISTORY OF ALZHEIMER'S DISEASE

Alzheimer's disease.

Speaking just those two words reliably sends shivers down the back of anyone over the age of 50. That fear is surely justified. Alzheimer's disease spares our body but robs us of our mind. It destroys our personality and, in doing so, destroys that special part of ourselves that makes us who we are. Among all human diseases it stands out from the rest by virtue of the fact that it is incredibly slow, totally relentless, and frighteningly common.

If we are going to conquer our fear of those two words—if we are going to conquer Alzheimer's disease—we need to understand the nature of the disease that we face and design our attack in the most strategic way possible. My goal for this book is to guide the reader through just such a strategy session. Along the way, we will learn a little biology and take a close look at how the biomedical research community has struggled to find treatments for Alzheimer's disease. As this book's title suggests, its central theme is that, sadly, we have made some big and costly mistakes in our search. Much of the book is devoted to laying bare these errors—but a book that only criticizes leaves the job half done. In the final chapters of the book, therefore, I offer a new vision of what Alzheimer's disease is and how this new model can be used to identify the disease, to study the disease, and to treat the disease. This first chapter, though, is devoted to the seemingly simple task of defining Alzheimer's.

DOROTHY'S STORY

Let's begin with the story of Dorothy. It is one story among millions of how a once vibrant personality literally slipped away from the body in which it lived. Dorothy's story was told to me by her daughter, now in her 70s. As I listened, I reconnected with the powerful emotions that drive me, and I believe other researchers, to devote their energies and passions toward finding a cure for this horrible disease. It also reminded me of the fear most people experience when they confront the question "Could this happen to me, and if it did, what would my family and I do?"

When I asked Dorothy's daughter what she understood Alzheimer's disease to be, she told me that she thought of it as a disease of the brain. Mostly though, she thought about it as a bringer of "heartache and stress"—the heartache of watching her mom be "sometimes there, sometimes not" and the stress of having to cope with her mother's ever-changing practical needs. This two-sided burden, she confessed, constantly sapped her own energy, patience, and strength.

Dorothy lived to be 96 years old. She was born in New Jersey but lived in Pennsylvania for most of her life. She was barely 5 feet tall but was a giant among her friends and family. Her husband had died when he was relatively young, but Dorothy never remarried. She had stayed active and very much enjoyed life. Her daughter was almost accusatory when she told me that her mom had avidly pursued all the activities that are supposed to protect you from Alzheimer's disease. She rarely missed doing the crossword puzzle, she walked everywhere and got plenty of exercise, she traveled and was very, very engaged with her friends. She was active in politics and civic duties. She loved to laugh and always appreciated a good joke. "People were drawn to her," her daughter told me.

Dorothy was relatively old when things started to change. I asked what the first inkling had been that something was changing—something beyond the normal slowing that one might expect with age. The first real hint, the daughter told me, came when Dorothy was in her late 80s. The hint came in a comment that a friend of her mother's made one day. The friend said she was concerned because when Dorothy would drive the two of them somewhere, she would get lost more and more frequently. Then there were a couple of very minor accidents. They were nothing, really, the friend assured the daughter, but she was getting anxious. The daughter was concerned but figured it was probably just the normal aging process taking its toll.

Nonetheless, the time seemed ripe to take action, so she and her sister convinced Dorothy to stop driving and to move from her home to an assisted living facility. Some of Dorothy's friends had already moved there, so it was not a very hard sell. At that point the daughter felt that she just needed someone to watch over Dorothy, but she marks this transfer from her longtime home to the assisted living facility as her and her sister's recognition that all was not right with Dorothy. The next turning point was marked by a pair of shoes. Apparently, Dorothy had recently bought a new pair of shoes to wear to a 90th birthday party that was being thrown in her honor. But when the daughter asked her if she was going to wear her new shoes to the party Dorothy responded, "No. I think I'll save them." "Well," said the daughter, "at your age, what in the world are you saving them for?" and they both had a good laugh. "She did wear them to the party," the daughter said, "And I have a picture of her walking up my driveway, grinning ear to ear and very happy about her new shoes."

A few weeks passed, and the daughter went to visit her mother. After a few pleasantries were exchanged, Dorothy got thoughtful and said, "You know, I think I'm going to return those shoes that I bought." The daughter was a bit taken aback. She said, "Well, you know, Mom, you really can't do that because you already wore them." "No, I didn't," said Dorothy. Even all these years later, telling this small part of Dorothy's story brought a tear to the daughter's eye. She told me, "This made it clear to me that there was a serious problem. I said to myself, 'This girl never forgets a pair of shoes. This is not good.'"

Dorothy was fine in the independent living situation and remained truly independent for another year and a half. Things seemed reasonably stable, but one day, the daughter said that Dorothy commented on how much she liked living in her new home. "And do you know it doesn't even cost me anything." After a pause in her narration, the daughter reminded me that Dorothy was a bookkeeper and all her life had paid her own bills and kept meticulous financial records. This was an almost unimaginable change. The daughter went to the office of the residence, and, sure enough, Dorothy had missed the last month's payment. So, the daughter talked to her sister and the two of them took over paying Dorothy's bills.

Things deteriorated faster after that. Dorothy forgot that she had a second bathroom in her apartment. Then she couldn't remember how to use the phone. Same with the clock radio. She would repeat herself more and more. Some of this, the daughter assured me, was probably normal aging, but it got worse and worse and worse. There were too many things to just pass off as simple aging. The daughter said that this was when they started going to doctors. A geriatrician gave Dorothy some tests and tried to explain their meaning. "Maybe I blocked a lot of this out," the daughter said. "To tell you the truth, I don't even remember clearly when it was decided that Mom had Alzheimer's. Maybe what they said was some kind of dementia . . . or Alzheimer's." I asked whether they had worked with the doctor to offer any sort of relief. "She might have been on a drug—Aricept maybe—for a short while. But Mom was never a big drug taker, and besides it didn't seem to do much for her . . . and the side effects . . . so we said let's not wear her down with all this."

Dorothy shortly progressed to a more intensive assisted living arrangement. She was OK there for a while, but the management was afraid she would wander in the middle of the night, so they moved her to a separate unit. That sent Dorothy into a tailspin. The daughter said that the move was unnecessary and frankly ridiculous. She had traveled extensively with her mother and knew her sleeping habits. As she told me, "My mom loved sleeping. When she went to sleep, she stayed asleep." But the facility was not interested in her travelogues and insisted that she allow them to move her mother to a locked unit. The daughter refused and spent days, taking time off from work, trying to find a place that would take her mom but not make her miserable as her faculties dwindled. "It was clear that the assisted living place really didn't care about her. They wanted to deal with her as a problem, not as a person." In the end, after a lot of searching, the original place called back and said that if the daughter would agree to putting an alarm on the apartment door at night, they would agree to take her back. The daughter agreed, and Dorothy stayed in the assisted living facility for another year. "And do you know how many times that fool alarm went off?" the daughter asked me. "Not once! I knew my mother. Alzheimer's or no she loved her sleep."

Walking was the next to go, and with time Dorothy started falling more. She started walking outside and getting lost unless she went with one of the people whom the sisters were paying to walk with her. But the solutions were increasingly short-term, and problems kept growing in number and intensity. The daughter felt the time had come. She suffered more days of lost work while she found a dedicated Alzheimer's facility with its own garden that Dorothy could wander around in.

Language was next. Dorothy's ability to remember and use words slowly but surely started going, especially, the daughter said, after she moved to the new unit. She came in as one of the more "high-performing" folks but quickly went downhill. And, perhaps as a result, the staff started doing less and less for her, which in turn only seemed to make things worse. Her motor abilities declined, and she had to use a wheelchair.

"She never forgot how to eat though," her daughter told me. "And the best thing was, she never forgot who I was. Most of her words were gone, but when I came in, she would always grin and raise her arms up and say, 'Yay.'" At this point, she liked simple pleasures, the daughter told me. They would go through art books together and look at the pictures or listen to Frank Sinatra together and sometimes sing along. The staff seemed to like Dorothy despite her limitations. She would occasionally have a nasty exchange with one the caregivers, but she was never truly aggressive with them or with other persons in the facility.

One day, after nearly two years in this memory unit, the daughter went to see her mother before leaving for a trip. "She was different," the daughter remembered. "It was still her, but she was not the same." A week later, Dorothy passed away at age 96 after having been in decline for almost eight years.

As you think about Dorothy, recognize that many of the details of her story are quite typical and will be recognizable to any person who has had firsthand experience with Alzheimer's disease. Some of the details of her case, however, are not at all what one might expect from a person with Alzheimer's. There is an important message in the "imperfection" of this example. I could have created a fictional person—a composite who captured all of the textbook phases and features of a typical course of Alzheimer's disease—but I decided not to. I came to this decision in part because a persistent feature of the disease is that it varies from person to person, and I have come to realize that this variation is an important part of what we need to deal with as we study this devastating human disease. I will explain the reasoning behind this decision more fully in chapter 12.

DEFINING DEMENTIA AND ALZHEIMER'S DISEASE

Precise or not, in many ways the story of Dorothy has within it all of the elements that we need to define Alzheimer's disease. We need to dig into this "natural history" lesson, therefore, and learn as much as we can about the causes of Alzheimer's disease and why it progresses in the slow and relentless way that it does. We must always keep the Dorothys of the world in the front of our minds as we push our research forward, but we cannot design pills or potions based only on stories. To find a cure, we will need to know more. We first need to define Alzheimer's disease.

That sounds simple, but unfortunately it is not. I spoke with many experts in preparation for writing this book. I began each interview with the same question: "Tell me, in your own words, what is Alzheimer's disease?" I assumed that this would be a simple way to get the conversation started and put my colleague at ease before we got to the more difficult questions. I was wrong. Going over my notes after many interviews, I realized that no two people had given me the same definition. This might sound weird to the average reader, but it touches on a core problem facing the Alzheimer's field: we actually don't agree on a definition of the disease we're working on. Different physicians and different researchers define it differently. Each person has their own perspective, and there are broad areas of overlap. Yet, as my interviewees' responses make clear, no single definition of Alzheimer's disease exists among the workers in the field. The differences are partly choices of emphasis. Some people based their definition on clinical symptoms, some on how the disease advanced, some on abnormal deposits in the brain, and some on genetics and family history. Most cited more than one of these features, and the priority assigned to each feature varied. That raises the question of how, without a clear definition, could any of these experts know when they had seen a real person with Alzheimer's disease. That is an excellent question, and we shall spend the rest of the book trying to answer it.

In fact, I will argue throughout the book that this lack of a single, accepted definition of Alzheimer's disease is an existential problem for the field. It holds back our progress more than any of the other problems we

will discuss. For now, however, let me offer the following working definition of Alzheimer's disease as our starting point. You will notice right away
that it's long and also that it's based only on symptoms. Later we will learn
what happens to the brain itself, but I suggest that what really matters to
us, as friends and family, is what is happening to our loved one:

Alzheimer's disease is a late-life disease that, over the course of many years,
destroys normal brain function in a progressive and irreversible fashion.
Throughout the advance of the disease, the person is largely unaware of the
dramatic nature of the changes that are happening to him or her. The inability to form new memories is one of the first hints that there is a problem. The
person then begins to lose his or her ability to perform complex tasks. As the
disease progresses, language skills and reasoning deteriorate as does the ability to make judgments. Personality changes such as depression and apathy set
in along with unexpected emotional outbursts, aggression, and agitation. The
ability to navigate worsens, leading to helpless wandering. For each of these
changes, the severity of the dysfunction increases with time. Through most of
the disease process the physical health of the person with Alzheimer's remains
strong despite the continuing mental deterioration. By the end stages, however,
the person becomes bedridden, incontinent, nonverbal, and nonresponsive.

I have tried to include the important symptoms of the disease—the loss
of memory, language, reasoning, and spatial orientation. I also included
the behavioral problems that are very much a part of the disease and
not a consequence—aggression, depression, and agitation. And I snuck
in one curious part of the disease that we haven't dealt with yet—the
strange lack of self-awareness on the part of the person with Alzheimer's.

A mentor of mine, David Geldmacher, would tell this story about new
patients. An older couple might come into his office for a routine neurological exam. Maybe the wife was experiencing numbness somewhere
or was having a problem with recurrent headaches; it didn't matter. He
would always ask, "Is there anything else?" If the wife said in response,
"Well, you know, Doc, I do notice that I'm having a lot of trouble remembering things these days. I'm worried that I might be getting Alzheimer's,"
he would assign a series of tests, but he would be pretty confident that the
problem was age, not Alzheimer's. But if, as they were leaving his office,
the husband took David aside and said, "You know, Doc, I'm a little worried about my wife. She seems to be having a bit of trouble remembering
things these days," David would schedule the same tests and would again
be pretty confident of the result. In this case, however, David knew with

reasonable confidence what the tests would show: Alzheimer's. It seems strange to those of us still blessed with all our faculties that we would not be aware of such major changes in our thinking. But this lack of self-awareness (known as anosognosia) is a feature of Alzheimer's disease, and so it has earned a place in our definition.

We now have a working definition of Alzheimer's disease, but we need to consider carefully the meaning of one more word, *dementia*, before we start to explore the history of Alzheimer's research. Dementia is actually a great word because its Latin origins almost define the word for us. The *de* of dementia signifies "apart." It has the same meaning here as it does in words like *deice* (as in deicing the wings of an airplane). The second part of the word, *mentia*, means mind. What a perfect and succinct way to describe what happens to the person with Alzheimer's. They go apart from their mind.

People will often ask what the difference is between Alzheimer's and dementia. The answer is simply that Alzheimer's disease is one type of dementia. Parkinson's disease in its later stages can also be accompanied by dementia; so can Huntington's disease. Vascular dementia, HIV-related dementia, Lewy body disease, frontotemporal dementia, and progressive supranuclear palsy are all recognized as forms of dementia. The important point is that dementia itself is a very broad category that describes an age-related loss of mental ability. Saying a person has "dementia" is similar to saying that a person has a "heart condition" or "cancer." It describes a state, not a specific disease. Designing effective treatment in each of these cases requires more than a broad definition.

Alzheimer's disease as currently defined is only one form—albeit the most common form—of dementia. We will come to learn in the next chapters how Alzheimer's disease was discovered, but we will also learn how its definition was intentionally inflated on at least three separate occasions to describe as much of the broad category of dementia as possible. When we look at the label *Alzheimer's*, we should already start asking ourselves what level of precision this label actually brings to the diagnosis of dementia. Is it as precise as saying "a form of breast cancer associated with a mutation in the BRCA1 gene"? Or is it more like just saying "breast cancer"? Or is it maybe even less precise than that?

For the most part, dementia is confined to older people, where it is referred to as *senile dementia*. Over 65 is a commonly used, if imprecise, boundary marker between young and old. Every once in a while, however, dementia strikes earlier in life. When it does, it is referred to as *presenile dementia*. This name is informative because it emphasizes that normal senile dementia is an expected part of the aging process. Presenile dementia is meant to describe the situation in which dementia sets in at a younger age—an age before you might normally expect to see it. We will learn about a very special case of presenile dementia in the next chapter.

ALZHEIMER'S DISEASE IN OUR WORLD

The story of Dorothy is a stern reminder of what a nasty disease Alzheimer's is. It affects the person with the disease and places a tremendous burden on the family and other caregivers. The details of Dorothy's story give a picture of one individual. To understand Alzheimer's disease as a public health problem, you need to multiply Dorothy's situation not by a hundred or by a thousand but by millions and tens of millions. The frightening truth is that Alzheimer's is alarmingly common. It accounts for 30 percent to 40 percent of all users of adult day care services or residential care facilities and almost 50 percent of the persons in nursing homes. In 2019, there were 5.8 million people with Alzheimer's disease in the United States alone, and perhaps as many as 50 million people worldwide. In the United States it is the sixth leading cause of death behind cancer, heart disease, accidents, and a few others (see table 1.1). By the time you reach the age of 85, the odds of your being affected by Alzheimer's disease are about 1 in 3.

Look at the table, though. At first glance, something doesn't make sense. If Alzheimer's is so common, why do nearly five times as many people die from cancer each year? And even more from heart disease? It would seem that if Alzheimer's were truly a common disease, it should kill lots more people every year. The explanation for this apparent paradox foreshadows some important facts about Alzheimer's disease.

First, Alzheimer's is slow; there is an average of ten years between its first diagnosis and death. Every case of Alzheimer's disease is different so remember we are talking about averages here. Indeed, Dorothy's eight-year progression was more rapid than most. Cancer, by contrast, is relatively fast. Once it is discovered, if left untreated, it can kill you in less than a year. That means that most of the cancer we are seeing right now

Table 1.1 Leading Causes of US Death—2017

Cause	Number of deaths
Heart disease	647,457
Cancer	599,108
Accidents	169,936
Chronic lower respiratory diseases	160,201
Stroke	146,383
Alzheimer's disease	121,404
Diabetes	83,564
Influenza and pneumonia	55,672
Nephritis etc.	50,633
Intentional self-harm (suicide)	47,173

first appeared only a short time ago; most of the Alzheimer's disease we are seeing has been with us for a decade or more. Thus, while Alzheimer's is common, it is agonizingly slow and the number of deaths per year is not as high as it is for other diseases.

Second, while we call the disease "common" what we really mean is "common among the elderly." Lots of people over the age of 80 have Alzheimer's, but very few (less than 5 percent) under the age of 65 have it. The situation with cancer is exactly the opposite. While the incidence of cancer does increase with age, if you make it to age 80 you are unlikely to get it. In fact, your risk of developing cancer starts to decline after the age of 50, as does your risk of dying from it (see figure 1.1). Because Alzheimer's is not common but rare in the fraction of the population under the age of 65, the overall number of people with the disease is smaller, and so is the death rate. This also has important implications for our understanding of the biological basis of Alzheimer's disease. It is a clear reminder of the central role that aging plays in its onset. That means if someone asks you what they can do to lower their risk of Alzheimer's, you already know what to tell them: Don't get old. We will come back to this issue again and again.

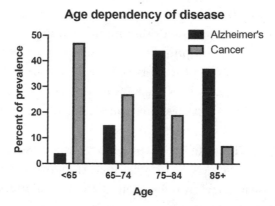

1.1 The prevalence of disease changes with age for both cancer (gray bars) and Alzheimer's disease (black bars). Cancer appears earlier in life and then reduces. Alzheimer's is rare before age 65 after which time it steadily increases.

Sources: Cancer data are from https://gis.cdc.gov/Cancer/USCS/DataViz.html; Alzheimer's data are from http://www.bocsci.com/tag/alzheimer-s-disease-389.html.

THE ECONOMIC IMPACT OF DEMENTIA

The slow chronic nature of a typical course of Alzheimer's disease means that it is a hugely expensive illness. The drain on the resources of our health care system is extraordinary. The estimated worldwide cost of dementia care in a single year is about a trillion US dollars. For the United States, in 2019, that number is estimated to be about $290 billion.[1] The exact numbers are difficult to calculate, but pause for a moment to contemplate their size. Let's use cancer again for a comparison. The total US cost of caring for all cancer patients is expected to be about $175 billion in 2020. Five times as many people die each year from cancer, yet Alzheimer's disease costs nearly two-thirds more. Once again, this seems like a contradiction, but the key lies in the fact that Alzheimer's is a slow, chronic illness. The cost of a single day's care is more expensive for a person with cancer, but the care is relatively short-term. With Alzheimer's disease the daily costs are much lower, but these lower costs need to be multiplied by the number of days in the ten or fifteen or even twenty years that the disease takes to run its course. The resulting financial burden that Alzheimer's disease places on our society is enormous.

On top of the huge price tag, it is worth remembering that the costs of this care are borne by more than the person with the disease. The health policy experts who analyze the situation have realized that the long, drawn-out course of the disease means that not only do the costs of direct medical care—in primary or secondary care facilities—need to be counted but so too do the costs of unpaid care provided by family and other caregivers, as well as the resources needed for support systems such as community care professionals and residential homeworkers. The indirect costs to an Alzheimer's caregiver such as a spouse or a child are hard to imagine. These are the people who tell the stories of Alzheimer's because these are the people who live that story every day. Academics try to put a dollar value on these indirect costs—lost wages for medical visits, lost career potential from having to go from full-time to part-time employment, and, most difficult of all, lost personal health of the caregivers due to the stress of dealing with a loved one with Alzheimer's.

You may be asking yourself, "Two hundred ninety billion dollars is a lot of money—who pays this enormous price tag?" The answer is not surprising but is memorable, nonetheless. Of the total $290 billion, it is estimated that Medicare and Medicaid pick up almost two-thirds. Another way to appreciate this direct burden on our health care system is to look at the average cost of a single patient. For a Medicare patient with Alzheimer's disease or some other dementia, the average amount paid per year by the federal government is almost $25,000. For a patient without Alzheimer's, that number is threefold less, about $7,500. The situation is even worse for Medicaid where the average benefit paid to a person with Alzheimer's is more than 20 times the benefit paid to a person without Alzheimer's or other dementias. As remarkable as these statistics may seem, some of the other numbers are equally memorable. Out-of-pocket expenses make up over 20 percent of $290 billion total—that's $63 billion out of the pockets of the person with Alzheimer's and their family. Think of all the things the families could have done with that many dollars. Sadly, private insurance covers only a tiny fraction of the cost. Basically, then, 90 percent of the care for your loved one with Alzheimer's disease is either on you or on your government.

WHO GETS ALZHEIMER'S DISEASE, AND CAN THE "WHO" TELL US ABOUT THE "WHY"?

The cost of Alzheimer's disease is very high, and yet all of us have skin in the game. If this were lung cancer, I would understand the logic of individuals who had never smoked a day in their life saying that they didn't see why they should help pay for other people's bad decisions. That argument doesn't work for Alzheimer's disease. If we ask who gets Alzheimer's, the answer is pretty much everyone. Or, at least, everyone who gets old is at substantial risk. That said, there are differences among different groups of people. These differences do not offer complete pro-tection to any person from any place on Earth. Each difference, however, offers a potential clue to the underlying biology of the disease.

For reasons that we partially understand, women are nearly twice as likely as men to develop Alzheimer's disease. At 45 years of age, a wom-an's lifetime risk of Alzheimer's disease is about 20 percent; the risk for a 45-year-old man is about 10 percent. By age 65, the lifetime risk for both sexes increases slightly, but the twofold elevated risk just for being female is still found. In part, this seemingly large gender difference is explained by the fact that women live longer than men. Life expectancy has been increasing around the world in the past few decades, but in virtually every country, women outlive men. In the United States, for example, the average life expectancy of a woman born in 2016 (81.1 years) is about three years longer than that of a man born in the same year (78.1 years).

Three years may not seem like a lot for an older person—that is, unless we're talking about Alzheimer's disease. Recall the importance of age as a risk factor. If we bore down into the numbers, what study after study has shown is that while the risk of Alzheimer's disease is very small when you are young, once you reach the age of 65, your risk doubles every five to six years for the rest of your life. Those numbers come together pretty well. Women live three years longer, Alzheimer's risk doubles every five years, and a woman's risk of Alzheimer's is about double that of a man. Once again, age matters.

Women's longer lives, however, may not explain everything about their increased risk. In a later chapter we will learn more about a variant of a gene called APOE that increases the risk of developing Alzheimer's disease. Curi-ously, a large study concluded that women who carry that variant were twice as likely to develop Alzheimer's, while men who carried the exact same vari-ant were only at a slightly increased risk. For now, the differences between the sexes is real, but the underlying reason for the difference is not clear.

What about geography? Same story. Globally, the rates of dementia and Alzheimer's disease show more similarities than differences. The world map in figure 1.2 shows that the rates of death from Alzheimer's disease do

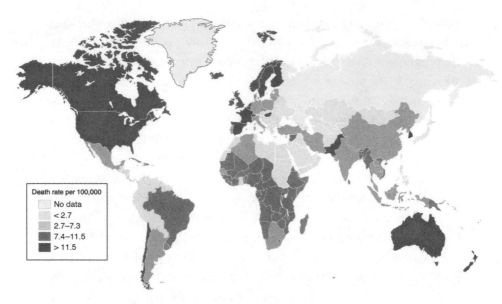

1.2 Global deaths attributed to dementia or Alzheimer's disease (data are from World Health Organization analysis in 2011).
Source: M. Yegambaram, B. Manivannan, T. G. Beach, and R. U. Halden, "Role of Environmental Contaminants in the Etiology of Alzheimer's Disease: A Review," *Current Alzheimer Research* 12 (2015): 116–146.

differ almost fourfold around the world.[2] And there are specific comparisons where the difference is even greater. Once again, however, overall life expectancy can explain a lot of these differences. Figure 1.3 is a life expectancy map, and you can see that the two maps look pretty similar. There are differences to be sure, sub-Saharan Africa being one example. But as with male/female differences, geographical differences are largely explained by age.

What about lifestyle? We have many interesting leads but few certain answers. Before we even get going, let me assure you that aluminum exposure (pots, pans, or cans) is not a risk factor. Nor are cell phones. With those largely debunked theories out of the way, epidemiologists have spent a lot of time pouring over the data from many different studies. In the process they were able to come up with some significant correlations.

When two things occur together over and over again, we say that they are correlated. Sometimes a correlation between two things is explained because one of them causes the other. Sometimes, but not always. For example, people who smoke tend to get lung cancer. Smoking, we say, is

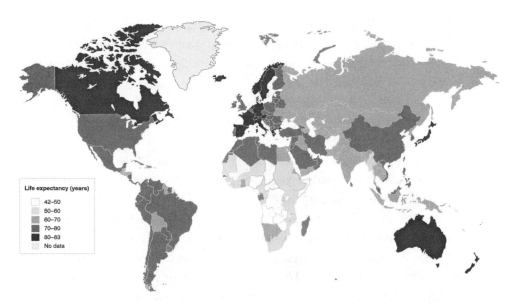

1.3 Global life expectancy (data are from World Health Organization analysis in 2011). *Source*: http://gamapserver.who.int/mapLibrary/Files/Maps/Global_LifeExpectancy_2008 .png.

correlated with lung cancer. That correlation had been known for many years, however, before the US surgeon general felt confident in saying that smoking *caused* lung cancer. The caution of the surgeon general was absolutely correct, despite the potential threat we now know that smoking poses to public health. To understand why, consider that people who smoke not only tend to have lung cancer but also tend to have ashtrays in their living rooms. Therefore, lung cancer is also correlated with ashtrays. But no one, least of all the surgeon general, is going to announce that ashtrays cause lung cancer. And no one is going to recommend that removing ashtrays, instead of cigarettes, from people's homes would be an effective public health strategy for preventing lung cancer.

Taking certain drugs is correlated with reduced Alzheimer's disease risk. Perhaps the strongest correlation is with the long-term use of nonsteroidal anti-inflammatory drugs (NSAIDs). Scientists found that the long-term use (two years or more) of high doses of NSAIDs was correlated with a strong reduction in the risk of Alzheimer's disease. The correlation was provocative because, as we'll learn, one of the features of the Alzheimer's disease brain is the presence of a slow, simmering inflammation. Unfortunately, the crucial next step in finding a treatment failed. Taking a group of people with Alzheimer's and randomly putting some on NSAIDs and giving some a sugar pill failed to prove that NSAIDs *caused* the reduced risk.

Persons taking the cholesterol-lowering drugs known as statins also seem to have a reduced risk of Alzheimer's disease. Once again, however, all attempts to prove, in a randomly selected group of people with Alzheimer's disease, that people taking statins could realize improved health compared with those who took placebo were unsuccessful. For now, both NSAIDs and statins are unproven in their ability to help with the progression of Alzheimer's disease. Other drugs have similarly failed, including antioxidants, vitamin E, ginkgo biloba, curcumin, and green tea. Yet for reasons we will delve into later, these correlations remain areas of interest and will be the subject of future studies. For now, however, they offer no magic pill.

Higher levels of education are correlated with reduced risk of Alzheimer's. People who never attended college are more likely to develop Alzheimer's than people who went on to higher education. But which way does the causality, if there is any, actually go? The most famous of the studies addressing this question was done out of the Rush Medical College of Northwestern University. Known as the Nun's Study, the researchers worked with the sisters in a midwestern convent to do an extraordinary longitudinal study. The sisters agreed to allow the Rush researchers to study every aspect of their life and follow them as they aged to see whether and when they developed Alzheimer's disease. The study was remarkable in many ways, in large part because the sisters lived their entire lives in the convent. This meant that many of the features of their day-to-day environment—food, exercise, sleeping habits, and others—were identical. One of the most surprising findings was based on going back and looking at the essays that the sisters had written as young women when they applied to enter the novitiate. What the researchers uncovered was that the "idea density" of the essays (as analyzed by a linguist) was negatively correlated with the risk of Alzheimer's disease 50-plus years later. A more complex wording in the essay meant a lower risk of Alzheimer's. A similar study was conducted in Scotland, where researchers tracked down people who had taken a nationwide IQ test in the 1940s and found that higher scores on the IQ test were correlated with a reduced risk of Alzheimer's disease.

Education and low Alzheimer's risk are correlated, but that's all we know. Maybe, if a brain is prone to develop Alzheimer's disease, the person is prone to have a reduced interest in schooling. Thus, despite the correlation, we cannot answer the question "Does education block Alzheimer's disease, or does Alzheimer's disease block education?" No definitive long-term study has yet been done to ask the question (as was done for NSAIDs) of whether, if we take a random group of people and give half of them education and give the other half a placebo, they develop Alzheimer's disease at the same rate.

Dietary choices are also correlated with modestly increasing or decreasing your risk of Alzheimer's disease. One well-studied example is a dietary regimen known as the Mediterranean diet—lots of fruits, vegetables, and cereals, moderate amounts of fish and dairy, but very low amounts of meat, sugar, and saturated fat. The reasons for the benefits of the Mediterranean diet are unknown in biological terms and are likely to be due to a combination of factors. One plausible connection is that people on a Mediterranean diet are at reduced risk for both Alzheimer's disease and for diabetes. Since a person with diabetes is about half again as likely as a person without diabetes to develop Alzheimer's, controlling blood sugar no doubt plays an important role in the correlation between the Mediterranean diet and reduced risk of Alzheimer's disease. As with education, the correlation with diet in general and diabetes in particular is persistent and reproducible but not yet understood in biological terms.

Diabetes is diagnosed clinically by regularly finding high levels of sugar in the blood, primarily a simple sugar known as glucose. Glucose itself, however, does not seem to be the connection to Alzheimer's. The problem seems to be with the levels of insulin in the blood. When your blood sugar levels go up (say, after you've eaten a big meal), your body responds by pumping more insulin into the blood. Insulin is a biological signal known as a hormone that tells the cells of your body to start taking up sugar. There are billions and billions of cells in your body, so, when they all start taking up sugar, the levels of glucose in your blood go down and things go back to normal. The problem in diabetes is that your blood sugar is always high, so your body is constantly pumping out insulin. That has the same effect as hearing a nonstop car alarm in the parking lot outside. You pay attention for a while, but eventually you realize that nothing is wrong. It's still annoying, but you basically stop listening. It's the same for your cells when insulin levels remain high for too long. Your cells just stop listening to the insulin signal; they have become insulin resistant. In my own lab we are working on this problem, and our early findings have begun to paint a clearer picture of how insulin resistance leads to dementia and Alzheimer's disease. Here again, no well-controlled clinical trial has been done. If we randomly divide a group of people and lower blood glucose (or insulin) in half of them, does that significantly reduce the risk of Alzheimer's disease in that group compared to the placebo group? For now, there is only correlation, not clear proof of causality.

There is one last study of environmental factors that we should consider. Researchers in Finland asked what the effect of controlling multiple environmental factors all at once would be. It was a relatively large study—1,200 participants—and it was long-term (two years). The Finnish Geriatric Intervention Study to Prevent Cognitive Impairment and Disability (FINGER) worked with lifestyle in a totally comprehensive way. Participants varied their nutrition, exercise, and social activity and did brain exercises. The researchers also used aggressive modifications of metabolic and cardiovascular factors. The results were encouraging. In just two years they observed a significant difference in the speed of mental processing, attention, and higher order thinking (known as executive functioning) in the experimental group. Memory also showed some improvement. There is work still to be done, and much of it is now underway. The study has to be repeated in other settings, the diversity of the trial population has to be increased, and the variables have to be separated and examined independently (there is evidence that just aggressively lowering blood pressure may be enough). Also, since this trial was done on persons without existing dementia, we need to find out whether the treatment works even after the clinical symptoms of Alzheimer's appear. Then too, we need to learn more about the underlying biology of how these factors interacted with age and disease to slow the emergence of dementia.

SYNOPSIS

Through the story of Dorothy, we have borne witness to the slow, relentless loss of brain function that we call Alzheimer's disease. In the process we have had our first encounter with a problem that will recur throughout these pages, namely, that there is less agreement than one might hope on an exact definition of the symptoms of Alzheimer's disease—who has it and who does not. Remember that Dorothy's daughter was not 100 percent certain nor apparently was her doctor that she had true Alzheimer's disease. This means the borders of the condition we are hoping to learn about are fuzzy, making our learning process more difficult and, as we will see, more frustrating. We learned the broader term dementia and how Alzheimer's disease represents the most common dementia, but not the only one. We have begun to appreciate the importance of age in triggering the onset of Alzheimer's disease and the other types of dementia. Finally, we surveyed the outside influences (environmental factors) for clues to the origins of Alzheimer's disease. Although there are features of our lives and life choices that are correlated with Alzheimer's disease, to date only the aggressive lowering of blood pressure has been shown to actually block the disease.

Knowing what you know so far, what would be your next steps in studying this human disease? If you were running a foundation with millions or even billions of dollars to give away and your goal was to conquer Alzheimer's disease, how would you spend your money? How much would you spend to make the lives of people like Dorothy better? How much on studies like FINGER to prevent people from getting the disease in the first place where the therapies were simple interventions such as exercise and other activities? How much would you want to focus on pharmacological approaches—pills and potions? There is more to learn, but even now it is worth keeping these questions in mind. We will learn about the history of Alzheimer's disease and how it is a textbook case of how not to study a human disease. But if we ask these questions properly, we might also learn how we can do better going forward.

2

A PHYSICIAN'S HISTORY
OF ALZHEIMER'S DISEASE

It is customary that if someone discovers a new disease, his or her name can be attached. Parkinson, Huntington, and Alzheimer were all physicians who described the maladies that will forever bear their names. The names of the people who actually had the disease and whom they studied to announce their finding are forgotten for the most part, their names buried deep in the scientific literature. To be sure, discoveries are important, and so are the people who make them. But naming a disease after a physician seems to de-emphasize the persons who actually suffer with the condition. Naming a disease after its discoverer also encourages all manner of nonscientific behavior whose goal is not to advance understanding but simply to pump up the importance of a disease that bears one's name. What the history of Alzheimer's disease teaches us is that these types of nonscientific marketing efforts can be deployed even if it is not one's own name on the disease. In point of fact, Alzheimer himself was not the one who coined the term Alzheimer's disease.

The signature case was a German woman named Auguste whose mental status began to deteriorate to the extent that her husband, a farmer, brought her to a psychiatry clinic in Frankfurt. Auguste had become forgetful and belligerent and required nearly constant care to the point where the farmer could no longer provide it for her. As is traditional, the scientific literature until very recently referred to Auguste merely by her first name plus the initial of her last name—Auguste D. The family name was withheld to preserve the family's anonymity. She was seen at the Frankfurt psychiatric hospital in 1901, and her attending physician at the time was a young 30-something anatomist turned psychiatrist, Alois Alzheimer. Alzheimer came to this case from early interests in the structure of the brain, basically its anatomy and cellular structure. He had learned to appreciate the newfound power of using microscopes to look at the cells of the brain as part of his studies. This very structure-based view of brain function was the context he brought to his clinical experiences in psychiatry, a craft he learned under the guidance of Dr. Emil Sioli in Frankfurt-am-Main.

Alzheimer's notes from the first interactions he had with Auguste D. are quite precise. This entry was made shortly after her admission to the hospital on November 26, 1901.[1]

She sits on the bed with a helpless expression.
"What is your name?"
"Auguste."
"Last name?"
"Auguste."
"What is your husband's name?"
"Auguste, I think."
"Your husband?"
"Ah, my husband." (She looks as if she didn't understand the question.)
"Are you married?"
"To Auguste."
"Mrs. D.?"
"Yes, yes, Auguste, D."
"How long have you been here?" (She seems to be trying to remember.)
"Three weeks." (She had been admitted on November 25th, one day before.)
"What is this?" (I show her a pencil.)
"Pen."
A purse and key, diary, cigar are subsequently identified correctly. At lunch she eats cauliflower and pork. Asked what she is eating she answers, spinach. When she was chewing meat and asked what she was eating, she answered potatoes and then horseradish. When objects are shown to her, she does not remember after a short time which objects have been shown.

This, then, was clearly a person with an advanced dementia. Many of the symptoms also match those of our working definition of Alzheimer's disease—emotional outbursts, failure of short-term memory, difficulty with language, and so forth. At the time, however, she was simply described as suffering from a somewhat aggressive early onset case of dementia. She went on to live for several more years, never regaining any of her faculties, and died in the Frankfurt facility in April 1906. By this time, however, Dr. Alzheimer was long gone.

Though he enjoyed his practice and worked hard to understand his patient's problems, Alzheimer also sought to advance his career. Thus, when he was invited by the famous psychiatrist Emil Kraepelin to join a group in Heidelberg, he accepted and left Frankfurt in 1902. Kraepelin was then offered a prestigious new job, which he accepted, and he moved, along with several members of his team, including Alzheimer, to Munich. To keep his interest in anatomy alive, Alzheimer got himself assigned the task of setting up a modern histopathology laboratory in his new Munich home. With all of his administrative and clinical responsibilities, Alzheimer was finding himself with less and less time to devote to his academic pursuits. In April 1906, his old mentor Dr. Sioli sent word to Alzheimer that Auguste D. had died. Alzheimer was pleased to learn that Sioli had arranged for an autopsy and had sent tissue from the brain of the deceased woman to Munich for Alzheimer to examine. The next events in this saga changed the course of dementia history.

As Alzheimer set about to examine this new autopsy material from Sioli he decided to use some of the advanced silver staining methods that his colleague in Kraepelin's group, Franz Nissl, was using. He prepared the brain tissue according to these modern recipes and peered down his microscope at the brain cells of Auguste D. He immediately saw that this material did not look like normal brain tissue. Two features of the preparation in particular caught his eye.

The first he described in this way: "Throughout the whole cortex . . . one finds miliar foci, which are caused by deposition of a peculiar substance in the cortex." The "peculiar substance" we now know was a waxy form of aggregated protein known as amyloid. The "depositions" would come to be known as amyloid plaques. The second feature he described thusly: "very peculiar changes in the neurofibrils . . . only a tangle of fibrils indicates where a nerve cell had been previously located." These peculiar neurofibrils are actually aggregates of a different protein known as tau. We now refer to these aggregates as neurofibrillary tangles.

In this way the odd deposits now known as plaques and tangles became tightly linked to a specific form of dementia that would come to be known as Alzheimer's disease. Alzheimer made detailed notes on his discovery and took them to Kraepelin and the other members of the group. He was quite certain that the plaques and tangles were the explanation for the highly unusual behavior of Auguste D. Kraepelin apparently liked the idea enough that he suggested that Alzheimer present his findings at a meeting of German psychiatrists in the fall of 1906. Alzheimer agreed and went to Tübingen that fall to announce his discovery. The response from the assembled scientists was reportedly underwhelming. Nonetheless, Alzheimer soon published his finding before returning to his work in Munich. He remained convinced of his interpretation of the cause of Auguste D.'s dementia and later went on to describe a second case (Josef F.) who had a similar rapidly progressing dementia and whose brain also had the amyloid plaques but did not have evidence of neurofibrillary tangles.

The story might have ended here as two long-forgotten case studies gathering dust in the archives of medicine. Alzheimer's boss Kraepelin, however, had other ideas. He too was a believer in the idea that psychiatric disease was caused by changes in the physical structure of the brain. The novel plaques and tangles that Alzheimer had found in the brain of Auguste D. neatly fit that philosophy. Kraepelin was very well-known at the time, in part because he was the author of a widely used textbook, *Psychiatrie*. That put him in a far more powerful position and allowed him to champion his philosophy to a much larger audience.

Kraepelin would periodically update his textbook to include the latest findings (and maybe to sell more books), and, as luck would have it, at the time that Alzheimer published his case, Kraepelin was preparing the eighth edition. To add more support to his own philosophy of the brain, he decided to include the case of Auguste D. in his revision. One might imagine it felt awkward to include a simple case study in a widely used textbook. Kraepelin cleverly solved this problem by elevating the case of Auguste D. to the status of a disease. To do that, he needed to give the disease a proper name. Perhaps he was a generous man or perhaps he did not want his own name attached to this thinly supported new entity, but for whatever reason he named the disease after his more junior team member. He called it Alzheimer's disease, and he included this new condition in the 1910 edition of *Psychiatrie*.

This was a bold and almost reckless move that, in retrospect, had a huge and outsized influence on the field. In most cases, I like it when scientists are bold. Put up a clever argument, and let a smart and informed debate refine it or rebuke it. Either way, science advances. Reckless is not so good. A textbook paragraph is much weightier than the same paragraph in a journal article or a meeting presentation. Textbooks impart a feeling of permanence to an entry. Their contents are imbued with the unspoken assertion that they represent settled art and thus are not easily questioned. Putting Auguste's condition in a textbook as a new disease comes pretty close to reckless because, on the flimsiest of grounds, Kraepelin was trying to put the "Case Closed" stamp on this telling of what he called Alzheimer's disease. It was to be the first of three inflations in the definition of Alzheimer's disease.

Let's go back and consider Alzheimer's findings in the brain of Auguste D. Two unusual features occurred together. Unusual deposits, plaques and tangles, were correlated with a highly unusual dementia. One possibility to explain the presence of plaques and tangles in the brain of a person with dementia is the one Alzheimer and Kraepelin favored: the plaques and tangles caused the dementia. That fit well with their philosophies that the function of the brain was governed largely by its structure. It most likely explains why Alzheimer championed this first explanation and why Kraepelin was so eager to promote it. Even though it's only chapter 2, however, you should already be able to come up with a second possibility. Perhaps Auguste D.'s peculiar dementia caused the brain changes that led to plaques and tangles. The plaques didn't cause the disease; the disease caused the plaques. And lest you think this is just some silly exercise, this is precisely the possibility that many members of the scientific community came to adopt in the years following Alzheimer's announcement.

Not only that, given the facts we have before us so far there are two other possibilities that could conceivably explain the correlation of the dementia and the plaques. The first is that things are actually more complicated than a simple two-body problem—plaques and dementia. We could hypothesize that the main driver of the dementia is a problem we have yet to identify, a third body. This third factor causes both the dementia and the plaques and tangles. The plaques and the dementia tend to come together, but they are not related to each other in a direct or meaningful way.

A diagram might help explain this better (see figure 2.1). Plaques and tangles are the square labeled P/T, dementia is the splat labeled D, and the unknown third factor gets a star with an *x* inside. Things that are correlated with each other are put inside a dashed line. The arrows represent causality. In Alzheimer's model the arrow points from the plaques to the dementia. In the alternative model, everything is exactly the same except the arrow points the other way—from the dementia to the plaques. In our newest model that I've called the three-body model, the correlation between plaques and dementia is still there (the dashed line box), but the arrow between them is totally missing. The cause of both the plaques and the dementia is a completely independent *x*-factor.

The fourth possibility that I've listed is a new one, but a simple one: there might be no correlation at all. It's just coincidence that the two features appear together. The plaques and tangles were there in the brain of a person with dementia, but this was just a chance occurrence. To show this in the diagram, I've taken away not just the arrows of causality but also the dashed line box of correlation. Why would this even be a possibility? Because the report on Auguste D. that Alzheimer published in 1906 is what is known as a case study. This is a common and valuable bit of medical research that is used when a physician comes across a new and

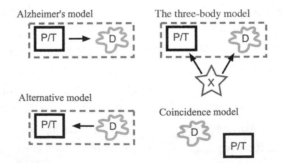

2.1 Models of correlation, causality and coincidence. P/T, plaques/tangles; D, dementia; X, unknown third factor. Groupings within dashed lines represent correlation; arrows represent causation.

noteworthy clinical case. There may be a new condition that no one has ever reported before (a 50-year-old grows a third leg), there may be a provocative correlation (visual hallucinations happen every time a person eats a brussels sprout), or there may be any number of other possibilities. Medical journals often have a special section entirely devoted to just these kinds of observations, but they are clearly recognized as "one-offs." Case studies are interesting and often noteworthy, but everyone understands that rigorous follow-up is needed before the new observations become clinically useful either to diagnosis or to treatment. To move beyond the curious to the scientific, a lot more work needs to be done. Rigorous follow-up means two important things: repetitions and controls.

It is often said that repetition is the gold standard of science, and that surely applies here. To prove that two things like plaques and dementia are strongly correlated, you need to see that they occur together in more than one or two individuals. Otherwise, you could find two things happening together totally by chance. Suppose you go to the supermarket to buy apples. They have green Granny Smith and red McIntosh apples on display. You pick up a red apple, and you notice it has a worm inside. On the basis of this one apple, after a bit of thought, you might get the idea that worms cause apples to turn red. We know that isn't true, but if all you had to go on was one red apple with a worm inside, it would be one of many possible explanations. At the same time, it's pretty easy to see that if you wanted to test your idea, you would need to look at more than one apple. This would be particularly important if other people in the supermarket told you that you were nuts. But if you looked at 10 red apples and 8 or 9 of them had worms, you would have a strong correlation that you could use to support your idea in the face of the criticism. Looking at 10 apples is called repetition, the gold standard of science. On the basis of the 10 observations you would be on more solid grounds in concluding that worms and red apples are significantly correlated. Are you ready to conclude that worms cause green apples to turn red? I made the apple example purposefully ridiculous because we all know it's not true. Worms have nothing to do with the color of the apple skin. But, so far, it would be logical and reasonable if you guessed that this was the case.

The apples also help us understand the idea of a control. Based on looking at 10 red apples, you have decided that the correlation of worms and red apples is a causal relationship. Your hypothesis is that worms make green apples turn red. If that's the case, then you might expect that the green apples would not have worms. This is called a control. You need to go back to the market, go to the apple display and pick up 10 green apples. If your hypothesis is correct, you should not find any worms in apples that are not red. Now imagine that you look at 10 green apples and 3 of them have worms. The first thing you would realize is that by looking only at the red apples, you had not been able to see the full picture of the apple/worm situation. Would you still stick to your idea that worms make apples turn red? Maybe, but probably not. This second observation—looking at 10 green apples—helped you "control" your thinking and take into account other things that might explain the worm/red apple correlation. That is the nature of a good control.

After finding all these worms in the supermarket apples, you would probably want to ask some questions. And since I made up this example, I will tell you what really happened. It turns out that the grocer was cutting corners to save money. He had gone to an apple orchard where the farmer let him pick up the dropped fruit for free. He still bought and paid for some fresh-picked apples, but most of his red and about a third of his green ones were "drops." In other words, they were older apples that had ripened and fallen to the ground where the worms had easy access to them. Worms don't cause apples to turn red or any other color. Worms-in-apples is caused by the age of the apples (and a penny-pinching grocer).

I accept that my example is a bit contrived, but I trust you see how this applies to the question of dementia and plaques. The case study of Auguste D. was the equivalent of looking at one red apple and finding a worm. Alzheimer for many good reasons was not able to repeat his observation on nine more Auguste D.'s, nor did he ever do his controls. What would have a control been in Alzheimer's case? His green apple equivalents would have been healthy people who did not have dementia. We know that he had looked at other brains, but they were mostly other cases with some sort of psychiatric problem. Having a true and fair control would have entailed looking in the brains of completely healthy people. In Alzheimer's defense, an autopsy is rarely done on patients who were healthy when they died. Thus, while the other brains he had looked at, his "controls," were most likely persons who had died with other nondementing illnesses, we can believe him when he tells us that the deposits in Auguste D. were unusual. We also know now that Auguste D.'s dementia had a particularly early age of onset. If the number of plaques in our brains increases with age, he might not have seen them in his other material. And given what we now know about the natural accumulation of plaques with age, we can be pretty comfortable in speculating that the other brains he looked at were from younger people and therefore less likely to have amyloid in them.

I've told the story of Alzheimer and Auguste D. in great detail because it tells us a lot about why the field is stuck and why a successful treatment for Alzheimer's has been so slow in coming. The original observation was an important case study, but it was elevated to the level of a disease for reasons that were strategic, not scientific. Both Kraepelin and Alzheimer were subscribers to a school of thought that held that the structure of the brain was the key to its function, and that if the structure became littered with abnormal deposits, its function would also become abnormal. Finding plaques and tangles in the brain of a person with dementia fit that idea, and putting it forward as a hypothesis was reasonable. From these origins, however, the two German psychiatrists inadvertently biased the thinking of subsequent generations of physicians and scientists. Their assertion that the correlation of plaques and dementia represented a causal relationship—plaques caused the dementia—has proven very hard to shake off. Given their evidence, however, it is about as defensible as asserting that worms cause apple skin to turn red.

I will close this chapter with a wonderful quote from a recounting of Alzheimer's life and work by Hippius and Neundorfer:[2]

However, . . . because this disease—presenile dementia with some unusual histological signs (plaques and neurofibrillary tangles)—was very rare, the name of Alois Alzheimer was almost forgotten for more than 50 years. During the last few decades, the situation has changed considerably.

This quote has two important parts to it. The first is that what Kraepelin and Alzheimer really defined was a specific type of dementia—"presenile dementia with some unusual histological signs." This is a perfectly legitimate classification, as legitimate as calling red apples with worms in them Alzheimer's apples. The second part of the quote that is noteworthy is that what Alzheimer and Kraepelin defined is in fact a very rare form of presenile dementia. In its original conceptualization, the term "Alzheimer's disease" applied to a dementia with "miliar" deposits that strikes early in life. It happens, but it's rare.

The last sentence in the Hippius and Neundorfer quote is also noteworthy because the situation has indeed "changed considerably." In fact, as we will see in the coming chapters, that sentence is definitely in the running for the understatement of the century. Is this truly how you would want to study a human disease as important as Alzheimer's? From his original obscurity, what were the forces that led to the extraordinary fame and notoriety that the name Alzheimer currently enjoys?

3

A SCIENTIST'S HISTORY
OF ALZHEIMER'S DISEASE

The role of a scientist in the study of a human condition such as Alzheimer's disease is different from that of a physician. Actually, to be more precise, the role of science in the study of a human disease is different from the role of medicine. When people walk into a medical doctor's office with memory loss and behavioral changes, they may be willing to wait for the results of this or that laboratory test, but their patience is not endless. They are worried, and they want answers right away—no ifs, ands, buts, or maybes. This is the challenge of medicine. The physician, as my doctor friend Jeff once remarked, needs to live by the axiom "Sometimes wrong, never in doubt." Make the best decision you can with the information you have, make it now, and move on.

A scientist, on the other hand, thrives on ambiguity and uncertainty because it is often in those fuzzy areas of a field that the newest and most exciting scientific findings are to be made. This is the realm of science and contrasts pretty dramatically with the situation in medicine. Not surprisingly, the contrast can generate tension among those of us in biomedical research. To understand how different these approaches are, consider that if a person with memory loss and behavioral changes were to make a wrong turn and end up in a scientist's office by mistake (we all wear white coats after all), the scientist would be likely to respond to the description of symptoms by saying "Gee. That's a really interesting problem you've got there. Give me 20 years to study it, and I'll get back to you." The person would be out the door in microseconds.

We will come back to this tension again and again. In its starkest form, the contrast looks like this:

Medicine: "We have millions of persons with Alzheimer's, and the public expects a cure. We need to do something right now, so let's take what we know, make our best guess, and start human clinical trials immediately."

Science: "We have nothing but the barest outline of the biological basis of Alzheimer's disease. We need a lot more basic science data before we can even begin to take an intelligent shot on goal. Human trials will have to wait."

This is a bottom-line tension that never goes away. Most of us in the field are not at either one of these extremes, although we may swing toward one pole or the other over time. In reality, all of us understand that both positions have merit.

Keeping this tension in mind, consider that our descriptions of Alzheimer's disease to this point have been from the perspective of the layperson or the physician. It's time to look at Alzheimer's again, but this time let us move beyond a mere description of the clinical symptoms and begin to probe our understanding of its biological basis. To do this, I am going to follow a rule that I learned very early in my scientific career. When you tell the story of a new discovery, it is almost always wrong to tell it historically. The process of discovery is messy at best. The pieces of the puzzle almost never fit together in the order in which you find them. Once the puzzle is finished, to try to describe the picture that emerges by telling how you fit each piece into place as you picked it up is confusing. It is almost never the most logical way to tell the story. What follows, then, is what I hope is a logical picture of the science of Alzheimer's disease rather than a historical account of how the puzzle pieces were assembled.

EARLY- VERSUS LATE-ONSET ALZHEIMER'S DISEASE

We learned in chapter 1 that most Alzheimer's disease first appears after a person has reached the age of 65. It is rare that symptoms start before then, and when they do, they are viewed as a special form of the disease.

This early-onset form tends to be more aggressive and, most importantly, it usually runs in families. For this reason, early-onset Alzheimer's disease is also known as familial Alzheimer's disease. We will learn later about why it runs in families, but for now we can just say that this is a rare genetic form of Alzheimer's.

The vast majority of Alzheimer's disease strikes its victim after the individual turns 65. This late-onset form of Alzheimer's disease is also known as sporadic Alzheimer's. And by the way, when doctors use the word sporadic to describe a disease, they're using doctor speak to tell people that they have no idea what's causing the problem. That's too bad because the sporadic form of Alzheimer's accounts for about 95 percent of the total number of cases. Despite these high numbers, for reasons we will get to later, most of the research in the field has been carried out on early-onset familial Alzheimer's disease even though it is sporadic Alzheimer's that is by far the bigger public health problem. Of course, you might want to ask, "If the two types of Alzheimer's are both Alzheimer's, then isn't it OK to study either one?" We'll see about that.

THE GENETICS OF SPORADIC ALZHEIMER'S DISEASE

It is hard to overstate the impact that the science of genetics has had on the biological sciences. In what we might call the old days (meaning before the mid-1980s) doing genetics usually meant mating males and females and seeing what happened to the parents' traits once they were passed to their children. This was slow and cumbersome and difficult to apply to humans. The situation was transformed, however, after the extraordinary advances of molecular biology were applied to the study of genetics. Using techniques with bizarre names such as *restriction fragment length polymorphism* and *single nucleotide polymorphism* (abbreviated SNP and pronounced "snip"), and using genome-wide association studies, we have gone from slow-and-cumbersome to fast-and-powerful ways to ask questions about the genes responsible for a given condition such as Alzheimer's. The changes have truly altered the face of biology, but this is neither the time nor the place to indulge in a primer on genetic techniques. Let's consider instead why you might want to study genetics in the first place if your goal is to understand the biology of Alzheimer's disease.

Our genetic code is like the blueprint of our biology. The entire blueprint is written in a simple molecular code that is faithfully reproduced in the DNA of every cell of our body. It is a chemical set of instructions about how to build biological things. Does your cell want to make insulin? The instructions for how to do that are written in code in the DNA of the cell's nucleus. Does one of your nerve cells want to make an Important Brain Protein? No problem. The instructions are in the DNA of its nucleus. And the instructions are reproduced more or less identically in the nucleus of every cell. Because all cells are working with the same set of instructions, the blueprints are even more impressive. Nature has figured out that it would be wasteful, if not outright dangerous, to get sloppy and maybe make the Important Brain Protein (let's abbreviate it IBP) in the liver, or an important liver protein in the brain. To guard against this, a lot of the DNA blueprint is taken up with information about how to regulate where and when things are made.

The DNA blueprint does not magically appear out of nowhere. It is given to us by our parents, each of whom gives us a full set. One set comes from Mom, one set comes from Dad, and this is how inheritance works. Mom's and Dad's instructions get together to give us our own unique set of blueprints. The entire DNA blueprint is known as our genome; one page from that blueprint is a single gene.

Great. But what does this have to do with Alzheimer's disease? Consider what would happen if Mom forgets to give us the gene for the IBP. Not a big problem usually, because we can always rely on the copy of the IBP gene we got in the blueprints from Dad. But what if Dad forgets too? We've got the same genome in every cell, but now, in every cell, the gene for IBP is totally missing. That means that not a single cell in our body knows how to make this critical brain protein.

Now, this might be a problem, or it might not. Our bodies are pretty good at figuring out work-arounds for lots of genetic problems like this. But in biology as in life, a work-around is never as good as the real thing. This is a particular problem for a protein that might serve as a part of our body's defenses against Alzheimer's disease. Going without IBP might not matter when we are young, with a risk of Alzheimer's disease that is close to zero. But as we get older and older, our not having IBP might find us having an increased risk of Alzheimer's disease.

That would be bad for us. It would be very useful for scientists, however, if they could turn around and replicate this correlation in lots of people. They might find that people who were more likely to get Alzheimer's disease were more likely to be missing the IBP page of their blueprints. If scientists could replicate this correlation enough times, it would give them an important clue about how to fight Alzheimer's. In this simple example, we might imagine that they could look for a way to supply extra amounts of IBP to older people. That way, even if these older folks had perfectly good copies of the gene in their own genome but weren't using it enough, adding extra protein might give them a boost and help them defend against Alzheimer's. That is the basic logic behind using genetics as a tool to study this or any disease—find the missing or damaged genes that are correlated with disease, and then restore working copies of the gene or help people find a work-around.

Yet this simple description barely scratches the surface of the power that genetics has brought to aid in the process of biological discovery. It will be useful, however, to explain one more important point. Notice that missing the IBP gene doesn't cause Alzheimer's disease in all people. If it did, we would call that gene a *disease gene*. What we said in this case was that people who were missing this gene were only more likely to get Alzheimer's. That type of gene is known as a *risk-factor gene*; it doesn't cause Alzheimer's, but it raises people's risk of getting it.

We know of three separate disease genes that cause familial Alzheimer's disease. We will talk more about these three later. In contrast to this situation, sporadic Alzheimer's disease has no known disease genes. Despite lots of looking, no scientist has found a gene that causes sporadic Alzheimer's in people who get it from their parents. That is probably telling us something important, but we can only speculate as to what it might be. In the meantime, scanning the entire genome for genes that are associated with a larger than normal risk of developing Alzheimer's, scientists have uncovered 29 risk-factor genes to date. That sounds like a big help toward our goal of finding a treatment, so let's take a look at this new treasure trove of information. Fasten your seatbelts. Here, in alphabetical order, is a list of the risk-factor genes that have been identified: ABCA7, ACE2, ADAM10, ADAMTS1, APOE, BIN1, CASS4, CD2AP, CELF1, CLU, CR1, DSG2, EPHA1, FERMT2, HLA-DRB1, HLA-DRB5, INPP5D, IQCK, MAPT,

MEF2C, MS4A6A–MS4A4E, NME8, PICALM, PLD3, PTK2B, SLC24H4-RIN3, SORL1, TREM2, and ZCWPW1.

If that seems to you to be more like alphabet soup than clues to the biology of Alzheimer's disease, you're not alone, but be patient. Let's break that list down into bite-size portions because there are indeed clues in that list.

When scientists look for genes that are correlated with a disease such as Alzheimer's, they very often use specialized genetic markers such as SNPs. The details are not important, but the power of the SNP markers is that they are found in millions of precisely identified spots all over our genome. At every spot, each local SNP can come in one of two "flavors." If a disease correlates with only one of the two SNP flavors, a scientist has a clue: that spot in the genome has some feature that alters disease risk. The scientist immediately starts looking for the gene that is located closest to that SNP and asks how a change in that gene might lead to Alzheimer's disease. Sometimes, the answer is clear; most of the time it is not, as is the case for the genes in the alphabet soup that I served up a couple of paragraphs ago. The answer is not clear, but there are many hints that are being followed up.

One big clue is where in the genome the SNPs lie. When scientists look for the spots with the highest correlation specifically to Alzheimer's disease, they find that the rogue SNPs tend to be in regions of our genome that tell us where and when to make something rather than how.[1] They are linked to a gene, but only because they are really close to it. This is an important clue because it points us to the idea that the problem in Alzheimer's disease is not how to build a biological something (like our aforementioned IBP). The problem is how much of it to make, or when to make it, or where to make it. In genetic speak we say that the problem is gene regulation, not gene structure.

This in an important difference because even in an era of advanced genetic engineering, it is still nearly impossible to correct the DNA code if we have to do it in all or even most of our cells. We can, however, affect gene regulation. In fact, you already know how to do it yourself if you remember the insulin story. Recall that when our blood sugar goes up, our cells make more insulin. That's an example of gene regulation. In a healthy person, if you want more insulin, all you need to do is raise the level of sugar in the blood—and all you need to do to raise your blood sugar is eat a big meal. Congratulations! By eating that candy bar, you just regulated your own insulin gene. I have oversimplified this example, but the basic point is accurate. We can often find ways to regulate where and when and how much our genes work by what we might call environmental changes. Remember the effectiveness of the Mediterranean diet? You can eat more, eat less, or eat different (especially if we include drugs as one of the ways in which you can eat different). Then, if we could learn to properly regulate the 29 genes in the alphabet soup, we might be able to make serious inroads into preventing or at least slowing the progress of Alzheimer's disease. If we are reading the genetic clues correctly, we may be able to manipulate their patterns of expression by environmental manipulations. We will still need lots of additional information, but this is a hopeful bit of logic.

The genetic help doesn't stop there either. Our body's systems are all interconnected. Our hearts don't do us much good if our brains have died, and our brains don't help us much if our heart has died. It's exactly the same with our genes. No gene works in isolation; they work in networks. Using the insulin gene again as an example, it turns on and makes more insulin if our blood sugar rises. But how does a gene in our genome know what's going on in our blood? The answer is that there are sensors made by other genes and relay switches made by still other genes. In the end, even to do something as simple as turn on the insulin gene in response to rising blood sugar, it takes a molecular village. Modern molecular biologists have organized and annotated these coordinated networks and assigned them to different broad cellular activities. For example, there might be an insulin signaling network. That is an enormous help if you're looking to find out what causes a complex disease like Alzheimer's. If there are multiple genes that alter the risk of a disease, even just a little bit, we can see whether these genes all come from one or a few of these identified networks. In the case of Alzheimer's disease, that is exactly what happens. Let's take a fresh look at our bowl of alphabet soup and see if we can find the network patterns. We can then take the rest of the known biology of Alzheimer's disease and see if the two pictures match up.

INFLAMMATION OF THE BRAIN AS A CAUSE
OF ALZHEIMER'S DISEASE

One of the gene networks with a strong and significant correlation to Alzheimer's disease is a network that is used by our genome in the process of inflammation. Of the 29 genes listed as associated with sporadic Alzheimer's disease, a group of them belong to networks that are associated with the immune system. This is a clue that inflammation is a key feature of the Alzheimer's disease process. Normally, inflammation is good for you. If you get a cut on your skin and it gets infected, an entire menagerie of immune cells cluster in the area and collaborate with one another to get rid of the infection. As these immune cells do their work, the area of the cut gets red and swells and gets a bit warm to the touch. All of these things are part of the inflammatory response, and they help our bodies get rid of the infection on our skin.

The problem is that sometimes our immune systems get all worked up over nothing. There is no infection, but the immune cells behave as if there were. This is the situation in rheumatoid arthritis, for example. In a nasty case of mistaken identity, our immune cells mistake our joints for hordes of invading bacteria or viruses and rev up their inflammatory responses to get rid of them. That turns out to be a bad call because it means that our immune system is attacking ourselves. The doctor who sees a person with arthritis will recognize this and often will suggest taking anti-inflammatory drugs. Although steroids are the most powerful drug of this type, the side effects can be quite serious, and since the person with arthritis will need to take the drug for a long period of time, steroids are not used very often. Instead, clinicians tend to prescribe a class of drugs called NSAIDs. These include such familiar drugs as aspirin or ibuprofen or naproxen. All of them have anti-inflammatory actions.

Does the evidence for the involvement of the gene network of the immune system give us a clue to the origins of Alzheimer's disease? Indeed, it does. It fits with an independent set of observations showing that there is a continuous low-grade inflammation in the Alzheimer's disease brain. The evidence for this was already pretty strong, and the genetic evidence made the case for an inflammation-based theory of Alzheimer's even stronger.

When people worked to extend Alois Alzheimer's microscope studies, they looked at more and more brains from persons who had died with dementia. As they did this, they noticed that a special brain cell known as the microglia cell was misbehaving in ways that looked like an inflammatory response. They knew what brain inflammation looked like from having looked at other brains that they knew were inflamed. The misbehaving microglial cells are the brain's version of the macrophages of our blood, and macrophages are one of the cells that cause the skin infection to get swollen and red and warm to the touch. Getting swollen, red, and warm is good for our skin if we want to stop an infection. It is not good for our brains. This is especially true because inflammation also delivers an entire cocktail of poisons to the area surrounding the infection. Our immune cells use these other compounds, called cytokines, to help kill the invaders and to signal to each other that an invasion of germs has begun. Scientists now know that the bigger problem for the Alzheimer's brain is this cocktail of poisons. Compounding the entire problem still further for our brains is that the microglia don't seem to know when to quit. In our skin, when the immune system triumphs over the invading bacteria, it shuts off. The redness goes away, the skin cools, and the swelling goes down. But in Alzheimer's disease there is no real enemy to conquer. There are no hordes of invading bacteria to kill. As a result, there is no signal to tell our microglia that they should shut off the immune response—and so they don't.

This cellular evidence of inflammation was an early clue to an out-of-control brain immune system as a key feature of the advance of Alzheimer's disease. When the genetic data came along, therefore, they fit very well with that earlier set of data. Even better, there was a third independent set of data that was pointing in the same direction. That a rogue immune response was a major culprit in sporadic Alzheimer's emerged from studies that were done to find associations (correlations) between Alzheimer's disease and various lifestyle factors. One area that received special attention were studies of people who took anti-inflammatory drugs. The idea, of course, was that if inflammation is a trigger for Alzheimer's, then if we were to take drugs to reduce inflammation, we might reduce Alzheimer's disease too.

How would you do such a study? You could recruit a lot of people, split them randomly into two groups, and give one group a regular dose of NSAID. The other group would get a fake pill (a placebo) and would serve as a control group. If you began with people who already had signs of Alzheimer's to see if you could stop its progression, you would be doing what is known as a therapeutic trial. You would be testing a therapy to treat an ongoing disease. But you could also start your study with two larger groups of people whom you recruit before any of them had any signs dementia. Again, one group would get the NSAID; the other would be a control. You would have to wait a lot longer till the natural course of events led to Alzheimer's disease appearing in the groups, but the goal would be to see if the NSAID group ended up with less dementia than the group on the placebo. This type of trial is known as a prevention trial—preventing a disease from ever starting.

There is actually another way to do a study like this. You can cheat: you could find somebody else's data and reanalyze them. OK. That's not really cheating. In fact, it's the work of an entire branch of science called epidemiology. To do an epidemiological study of the inflammation question, you would find a completed study where people taking anti-inflammatories had been analyzed for some reason totally unrelated to Alzheimer's disease. If you can then measure the rate at which dementia appeared (either from the original record or by recontacting the people to do a follow-up), you could reanalyze the data to see if the anti-inflammatory group had a reduced risk of dementia.

You might already be thinking about how to do this because a few paragraphs ago I told you that doctors who had seen persons with rheumatoid arthritis often gave them large doses of anti-inflammatory drugs. Sure enough, when epidemiologists analyzed the people in the arthritis trials, they found that those who had taken very high doses of NSAIDs for long periods of time had a significantly reduced risk of Alzheimer's disease. Other studies followed, and most replicated the original findings. High doses of NSAIDs lowered the risk of Alzheimer's by about half.[2] These replication studies also showed that it was not the arthritis itself that lowered Alzheimer's risk, it was the NSAID. Not only that, but when epidemiologists started comparing the various studies that had been done, they found that different types of NSAIDs varied in their effectiveness. Aspirin, for example, seemed to have little effect. Ibuprofen (Advil) and naproxen (Aleve), however, showed positive effects. Acetaminophen (the active ingredient in drugs like Tylenol) was tried even though it is not officially an NSAID and also proved less effective.

Two points need to be made right away. First and most important do NOT put down the book, drive to the pharmacy, stock up on ibuprofen, and start popping pills. The same goes for all of the other NSAIDs. The likely side effects of the massive doses you would need to take include serious problems with your kidneys, liver, and intestines, and the possible benefit is not worth the risk. The second point is that despite the seemingly good news from the epidemiologists, there is bad news to go with it. Spurred by the potential for a powerful new treatment, clinical trials of NSAIDs have been done. These were therapeutic trials that started with people who already had symptoms of dementia. In every case, across different NSAIDs (including a trial of an anti-inflammatory steroid), the group on the drug failed to show a clinically meaningful effect. I feel bad having to relate this result. The need for a treatment for Alzheimer's is so critical. The promise of anti-inflammatory approaches supported by three independent bodies of data—the genetic data, the microscopic data, and the epidemiological data—seems so huge. By all accounts, the trials should have worked. We seem to have made a start, but unfortunately, it's back to the drawing boards for now.

POOR FAT MANAGEMENT AS A CAUSE OF ALZHEIMER'S DISEASE

A second look at the list of genes that have been identified as significant Alzheimer's disease risk-factor genes reveals another important clue as to what might cause Alzheimer's disease. In the network analysis, another group of the 29 genes was identified as involved in one way or another with the ways in which we handle fat. Reading that, you might worry that we are going to talk about the ratio of cheeseburgers to kale in your personal diet, but fear not. We are not talking about how we handle fats in our diets. We are about to explore how our cells handle fat in their molecular diets.

First of all, it's important to note that there is overlap in the networks; some of the genes in the immune system network are also here in the fat network. It is a restatement and an extension of the idea that no gene is an island. No network is an island either. Genes work together to form networks, and networks work together to perform functions. One smart way to coordinate internetwork activity is to have the same pieces perform similar functions in different tasks.

Fats and oils (also known as lipids) turn out to be a serious problem for our cells. That's because we are mostly water—70 percent of our body mass—and so are our cells. Things that dissolve in water move freely in the cell, but things like fats and oils do not. They just sit there. What's a cell to do? Well, for one thing, the cell makes use of the fact that water hates fat and uses lipids to make containers to hold water and watery solutions. In fact, the entire cell is basically just a big water balloon with the skin of the balloon being the fatty outer membrane of the cell. Inside the cell is an entire collection of smaller balloons (the official trade name for these smaller balloons is vesicles). These are like the mason jars and Tupperware containers of our cells. They hold things that would be dangerous or sloppy if they were floating around randomly inside the cell. So, cells make use of fats to form barriers to keep these solutions in their place. That, however, only helps with part of the problem. The other problem fats and oils pose for the cell is that they need to be moved around from one place to another—within the cell and between cells. The cell solves this problem by having genes designed to make special proteins that serve as lipid carriers—either active carriers like pumps or passive ones that allow the transport of even the oiliest lipid through the watery environment of our cells. The passive transport proteins are nano-engineered to hold onto the fats even while they themselves remain able to move around freely in water. Clever cell.

So, what does this all this fat have to do with Alzheimer's disease? The honest answer is we're not sure. Remember the genetics is only a clue to the fact that the fat-handling network is off-kilter in Alzheimer's disease. Scientists have several theories as to why that might be, but none of them has reached a high degree of certainty just yet. The risk-factor gene that is the poster child for this line of thinking is the gene encoding a protein known as apolipoprotein E, APOE for short. If you go back and check the alphabet soup of genes, you'll find that it is number five. It's number five in alphabetical order, but it is by far the number one risk-factor gene for sporadic Alzheimer's disease. APOE is a lipid-carrying protein that is found in the brain but also in the blood. Lipids are everywhere, after all.

The variety of the APOE gene known as APOE4 turns out to be a major Alzheimer's disease risk-factor gene. There are two other variants that either have no effect on your Alzheimer's (APOE3) or may slightly protect you (APOE2). If you get an APOE4 gene variant from one parent, your risk of Alzheimer's goes up roughly five times. If you get an APOE4 gene variant from both parents, your risk goes up roughly ten times.

APOE is a pretty versatile lipid carrier. It can handle simple fats, and it can also carry cholesterol, which, like a fat, cannot be dissolved in water. The cardiologists were already after APOE and the APOE4 variant because it turns out to be a factor in heart disease. It was the cholesterol-carrier feature of APOE that caught the cardiologists' attention. APOE4 turns out to be less efficient at carrying cholesterol than the APOE3 variant. Perhaps, the reasoning goes, there is more unbound cholesterol bouncing around and able to start forming deposits on the walls of the blood vessels of the heart. The Alzheimer's connection is less certain. As we discussed before, what's good for your heart is good for your brain, so the connection could be as simple as that. The walls of the blood vessels of the brain also start accumulating cholesterol deposits, blocking blood flow and reducing brain health.

There are two other functions of APOE, however, that may be relevant to Alzheimer's disease. The first is that APOE can carry amyloid—the stuff that makes up the plaques that Alois Alzheimer described. The second is that APOE not only carries cholesterol in the blood but it also carries it from cell to cell in the brain. This is potentially relevant as there is a specialized fatty membrane called myelin that is full of cholesterol. As we age, the cells that make that myelin need to import more and more of their cholesterol from outside sources. More on that later.

APOE is only one of the genes identified in the fat management network. Others, such as ABCA7 and INPP5D are also involved in moving cholesterol around, in particular from the outside watery environment, across the fatty membrane of the cell, to the inside watery environment. The other genes identified play similar specialized roles. Their individual contributions to Alzheimer's disease are small. No one variant of any of these genes changes the risk by more than a factor of 2, most by less than 50 percent. But they are genetic clues that moving cellular fat is a part of what we need to take care of in our fight against Alzheimer's disease.

IMPROPER VESICLE MANAGEMENT AS A CAUSE OF ALZHEIMER'S DISEASE

Now we come back to the mason jars and Tupperware containers—the vesicles of a cell. There are six genes from the list of 29 that show up in the vesicle trafficking gene network (APOE, BIN1, CD2AP, PICALM, PLD3, and TREM2). Once again, the networks overlap each other. The APOE gene that we first met as a lipid carrier is certainly on the list, and that makes a certain amount of sense. Each of the vesicles is surrounded by its own lipid membrane, and if the lipids aren't right, the vesicle isn't right either. Note, too, that there is also overlap with the immune system network. That also makes sense because the cells of the immune system are, after all, cells. They need their vesicles to work properly just as much as a neuron. BIN1 is unique to this list, but its presence makes a certain intuitive sense. That's because one of its purposes is to bend membranes, and if you're going to build a vesicle—a tiny sphere of liquid surrounded by a fatty membrane skin—you are going to have to take a flat membrane and bend it.

Beyond these logical areas of overlap, however, the genetic identification of the vesicle network as a player in Alzheimer's disease is a clue that something else, something independent of fats and immune cells, is involved. One area where this may be especially true is in the process of garbage disposal. Like people, no cell is perfect. They make mistakes when they build things, and funny-shaped scraps and unfinished projects just get tossed into the cytoplasm (the cell's watery interior). Proteins can get messed up if they don't fold in just the right way. Other cellular elements like mitochondria get old and sometimes need to be "retired." The cell is a tidy critter though; it does its best to clean up after itself. It is also a master recycler and never throws away anything if it can be reused. One of its tricks is a specialized set of vesicles known as autophagosomes and lysosomes that are basically the cell's garbage disposal system. The details can be mind-numbing, but basically the cell has garbage detectors (the autophagosomes) that shrink-wrap cytoplasmic junk in a double membrane and ship it to the dip tank (the lysosome) where the junk is digested and dissolved by an impressive array of nasty proteins that are kept well apart from the rest of the cell by the fatty membrane of the lysosome vesicle. Once the digestion is complete, what's left is a solution that contains basic cellular building blocks that are returned to the cell's cytoplasm to be reused. The network analysis does not point to this garbage disposal system directly, but there are independent observations of the biology of the cells of the Alzheimer's disease brain that suggest there is an age-related loss of the ability of our cells to deal with garbage. Scientists see the garbage piling up both inside the cells where mitochondria are allowed to stay on well past their "retirement" and outside where things like amyloid plaques and neurofibrillary tangles start to accumulate.

In the end, though, the vesicle connection identified by the genetics does not point in any obvious way to a cause of Alzheimer's disease. There is good supporting biology, but not enough to make a convincing case that we drop everything and start improving the garbage collection in our cells and our bodies.

NONGENETIC CLUES TO THE CAUSE OF ALZHEIMER'S DISEASE

These types of studies should give you a practical sense of why genetics is such a powerful discovery tool in Alzheimer's disease research. With all of its power, however, genetics is neither fully comprehensive nor flawless. The best evidence for this is that there are piles and piles of nongenetic evidence that hint at other causes of Alzheimer's disease. These piles need to be taken quite seriously, even in the absence of clear changes in their genetic networks. For example, there is evidence from microscopic studies that a specialized cell structure known as a mitochondrion ceases to function properly in Alzheimer's disease. Mitochondria (that's the plural of mitochondrion) are the powerhouses of our cells. They are how we turn the food that we eat into useful chemical energy that our cells can use to do work. Based on the correlation between dementia and weird mitochondria, a school of thought has developed that it is this failure of energy production is one of the root causes of Alzheimer's disease.

Another, even larger pile of evidence concerns the role of calcium. Calcium is a metal, but a single atom of calcium is a powerful signaling force in biology. That's because many proteins are designed to bind a calcium atom, and when they do, their activity changes. The cell takes advantage of this regulatory power and adjusts the moment-to-moment levels of calcium with great precision—both inside and outside the cell membrane. The connection to Alzheimer's disease comes through the observation that the cell's ability to regulate calcium levels changes with age. This changes the activity of all of the calcium-binding proteins, and for reasons that we do not yet understand, these changes happen faster in the brain of someone with Alzheimer's disease. No experimental manipulations have yet been done to test whether restoring calcium levels would prevent Alzheimer's disease, so for now we must treat this as a clue, but not yet a path to therapy or prevention.

A more detailed stack of findings concerns the role of myelin. Myelin is an interesting part of our brain's biology. Most nerve cells communicate with other nerve cells using a type of brain electricity that relies on charged ions traveling across the nerve cell membrane. The business part of the communication is done along a long, thin cell process called an axon. You can think of an axon as similar to the wire in the walls of your home. The wire carries the electricity that powers the plug that lets you light your lamp or charge your phone. The analogy isn't perfect, but one way that a wire in your wall and an axon in your brain are similar is that they will both work better if they are surrounded by insulation. In the case of the lamp, it's the rubbery stuff around the metal wires that lets you touch them safely. In case of the nerve cell, it's a tightly wrapped jelly roll of fatty membrane called myelin that does the insulation. Axons with myelin work much more efficiently than axons without myelin. They move the current faster from point A to point B, and they do it more efficiently, so it takes much less energy. So why would myelin have anything to do with age or Alzheimer's disease?

There are many answers to that question, but one of the most compelling is that the parts of our brain that have the most myelin are the same parts that are hard hit by the Alzheimer's disease process. You may think your brain was done growing by the time you were 10 years old, or maybe 15, but you would be wrong. In fact, a human being continues to add myelin to some brain regions well into their 30s. And while it is stable for a few years after that, beginning in our mid-40s, we start to lose our myelin beginning in the very places where it was last added. Using modern imaging technology, scientists can see myelin in the living person and follow it over time. When they do this, they find that the amount of myelin in our brains is correlated with the strength of our mental abilities. This connection between the amount of myelin and our cognition is relevant to our discussion because the age-related loss of myelin happens faster in persons with Alzheimer's disease.

The connection of myelin to Alzheimer's disease, however, goes much deeper than that. The genetic analysis does not turn up a network specifically devoted to myelin, but many of the genes that appear as risk factors are quite relevant to the process of building it and maintaining it in the brain. For example, the cells that build the myelin wrapping of the axon (they're called oligodendrocytes since you asked) need to make an extraordinary amount of membrane to effectively insulate the axons. Membranes, as we now know, are made up largely of molecules of fat, and we have already seen that the genetics of Alzheimer's point to fat management as a deeply involved network. Even more intriguing, almost every membrane in a cell has at least one or two molecules of cholesterol embedded in with the other lipids. Cholesterol acts like a stiffening agent that makes the membrane a little less fluid, and for reasons related to their particular properties, myelin membranes have more cholesterol than just about any other membrane in our bodies. That fits well with the finding that a lot of the fat-management genes are related to cholesterol transport, including the APOE gene. As with the calcium hypothesis, there is no step-by-step pathway that can be drawn from myelin to Alzheimer's disease. But there are enough hints that it certainly seems like there should be more than just random correlation going on.

Mitochondria, calcium, and myelin are examples of the many biological, but nongenetic, clues to the causes of Alzheimer's disease. There is also a reasonable stack of evidence that there might be an infectious basis to Alzheimer's disease. The idea that a virus or a bacterium might be the trigger that starts a human brain on the path to dementia and Alzheimer's disease has long been discussed in the field. Though historically viewed as a fringe idea, recent work suggests that there is more here than might be imagined. Using sophisticated detection methods, scientists are able to find viral DNA and viral proteins in the brains of persons who died with Alzheimer's disease. Such evidence is much rarer in individuals who died with normal cognition. Because it has been on the fringe of the field, only a few researchers have actively followed these ideas.

Nonetheless, with increasing regularity reports have been appearing containing additional evidence for a correlation between an entire menu of different viruses or bacteria and the presence of Alzheimer's-like changes in the brains of experimental animals and humans.[3] These associations are found in every dimension. As mentioned above, viral DNA and protein are found in the Alzheimer's disease brain. More provocatively, when analyzed in fine detail, cells in the Alzheimer's disease brain are behaving at the molecular level as if they were fighting a viral infection. One epidemiological study even found a correlation between the persistent use of antiviral drugs and a reduced risk of developing dementia.[4] There is even provocative evidence from mice that if you administer certain bacteria to mice by way of a nose spray, they begin to develop the characteristic plaque deposits of an Alzheimer's disease brain.[5] Perhaps most intriguing of all are the arguments found in the scientific literature that the amyloid protein subunits—the building blocks of Alzheimer's amyloid plaques—have bacterial fighting properties. This opens up the possibility that the plaques are not only not a cause of Alzheimer's disease, they may actually represent evidence of the brain's effort to fight off an infection that could itself be the cause of Alzheimer's. If this sounds like a possible x-factor, it should. It's a perfect example of how dementia and plaques could be correlated with each other, but not because one of them caused the other.

The list of biological hints as to the origins of Alzheimer's disease is long, and not all entries will be considered here, but no list of these hints would be complete without considering what is known as the cholinergic hypothesis. The basis of this classic view of Alzheimer's disease is a small molecule, acetylcholine, that neurons use to talk to other cells. Neurons move information along their axons with the brain electricity that I described above. The information is really a code of "pings" that can come one by one or in a burst. The pings move along the axon quickly (and even faster along an axon with myelin), but when a packet of information, a ping, arrives at the end of the axon, it is passed to the next cell not with electricity but with chemistry. Each ping drives the specialized machinery at the end of the axon, known as a synapse, to give a little squirt of chemical. Ping-squirt, pause; ping-squirt, pause; ping,ping,ping-squirt,squirt,squirt. The next cell in the chain reads the squirts and turns them back into pings.

The chemicals that the neurons use in their squirts are called neu-rotransmitters (they "transmit" a ping from a neuron to the next cell), and one of those neurotransmitters is acetylcholine. The acetylcholine neu-rotransmitter got linked to Alzheimer's disease because two clever research-ers realized that when young people with healthy brains were given a drug that blocked acetylcholine from acting, they developed problems with their memory and even problems performing tasks that were more cogni-tive in nature.[6] The researchers made the connection to the changes seen in elderly people and then made the conceptual leap that losing acetyl-choline function might be a part of age-related memory loss and possibly Alzheimer's disease. They got close. So close that this insight forms the theoretical basis of the most commonly prescribed treatments for Alzhei-mer's disease: Aricept, Reminyl, and others. Close, but not close enough. That because, unfortunately, while the drugs work to improve memory, they only affect the symptoms. The disease process continues to advance even though some function can be restored. As a result, the effectiveness of the drug diminishes with time. This suggests that the real problem with Alzheimer's disease lies upstream of the acetylcholine system.

BACK TO PLAQUES AND TANGLES

All of these genetic and nongenetic hypotheses join hands with the origi-nal idea of Kraepelin and Alzheimer that the true smoking guns in the development of Alzheimer's disease are the plaques and the tangles. This telling of the disease has been embellished over the years to the point where many have suggested that it is solely the emergence of the "mil-iar deposits" and "peculiar changes in the neurofibrils" (the plaques and tangles) that are responsible for driving the dementia and behavioral changes that we recognize as Alzheimer's. Apart from the attraction of the structure/function correlation that so intrigued our two German psy-chiatrists, problems immediately arose in trying to attribute causality to the correlation. One early problem that was recognized by people explor-ing the distribution of the deposits in the brain is that their pattern does not very accurately reflect the pattern of functional problems. This is a particular problem for the plaques, whose density is greatest in areas of the brain that control functions that are not that closely related to the symptoms of Alzheimer's disease. Similarly, there are brain regions that are known to be deeply involved in the symptoms of Alzheimer's, and even lose a lot of neurons to cell death, but they do not have a significant number of plaques. The tangles correlate much better with function, but for reasons that we will come to understand in the next chapter, this stronger structure/function correlation has not been vigorously pursued.

In fact, virtually everything you read about in this chapter was left in the dust as the early 1990s ushered in an era wherein the amyloid plaques rocketed to the front of the line of probable causes of Alzheimer's disease. They hold this position to this day, and to truly understand how this has distorted our approach to Alzheimer's disease, we need to explore both the good reasons and bad ones for why that happened.

4

MYSTERY SOLVED! HOW FOUR DISCOVERIES TRANSFORMED AN ENTIRE FIELD

. . . and with the exciting new discoveries of the past few years I can confidently predict that we will have a cure for Alzheimer's disease within five years!

To be honest, that's not a precise quote, but it's close. I distinctly remember sitting in a darkened room among hundreds of other Alzheimer's scientists. We were listening to a keynote address from a very prominent Alzheimer's disease researcher who spoke words to this exact effect. The year was 1995.

Some simple arithmetic tells us that this was about 25 years ago, making it pretty apparent that the prediction was, shall we say, premature. So how did this very smart, very prominent researcher get it so wrong? And it was not just this one speaker. There were a lot of nodding heads in the audience, too. The rest of this book is an attempt to untangle the story of what happened, but my goal in this chapter is to try to recreate the atmosphere of those years. Though it may feel like just an exercise in the social history of science, to understand where we are today it is essential to recapture the enthusiasm and the optimism of the years between 1984 and 1999. The momentum that built up during those years, now seen 20 years distant in the rearview mirror of history, is a big part of the explanation of how we messed up.

Those 15 years were a time when the promise of genetics as a tool of discovery in the fight against human disease had never seemed to have more promise. Discovery after discovery seemed to push the frontiers of science ever further into the light. There was a near perfect congruence of biochemistry, microscopy, and genetics. Sure, there were concerns, but the flood of positive reinforcement fooled us into thinking that we had this dread disease by the jugular and were going to squeeze the life out of it. We weren't wrong to feel that way initially, hence this chapter. We were wrong because we couldn't let go of a bad idea once we learned more about it, hence the rest of the book.

UNMASKING THE PLAQUES

The "miliar deposits" that Alois Alzheimer found in Auguste D.'s brain continued to intrigue many a scientist. The plaques, as it turns out, weren't made of cholesterol—the yellowish plaques that clog the coronary arteries of a person with heart disease. They were made of the aforementioned waxy substance, amyloid. Amyloid happens when lots of copies of one small piece of a protein (called a peptide) stick to each other and form a nearly insoluble structure called a β-pleated sheet. Amyloid formed from this β-pleated sheet arrangement is a structure that is not unique to just one type of peptide. Pieces from over a dozen different proteins can make amyloid, so seeing an amyloid deposit anywhere in the body, including the brain, doesn't tell you much. It could have many different protein origins. Alzheimer's disease amyloid was of this type; it was made of protein, but that was all we knew. Alzheimer's researchers weren't satisfied with this ambiguity. Luckily, as the twentieth century drew to a close, the technology was becoming available that let us learn more and more of the molecular structure of things like the amyloid plaque of the Alzheimer's brain. Then, and still now, this seemed like an important problem to attack. The motivation was that if Alzheimer and Kraepelin were on to something with their ideas about the plaques causing the disease, then if we knew where the amyloid in those plaques came from, we could follow the trail back toward its beginning and hopefully divert the process from starting—a cure for Alzheimer's.

This line of thinking led George Glenner and Caine Wong, two scientists at the University of California at San Diego (UCSD) to try to isolate the plaques and see if they could figure out what they were made of. They started with a side observation of Alois Alzheimer's that the plaques were deposited not just in the substance of the brain but also in the blood vessels that serve to nourish it. These blood vessel plaques were thought at the time to be unique to Alzheimer's disease. In pursuing their research, they took advantage of the fact that some of the vessels that feed the brain run across its surface before they dive deep into the structure. These large vessels are also filled with amyloid in persons with Alzheimer's disease, but reportedly not in people without the disease. It was from this easily accessible source, therefore, that the first Alzheimer's disease amyloid was isolated and purified.

Looking back at the original description, it is easy to see how this entire enterprise could have gone wrong. The tissue that they used to get the amyloid was not from brain at all. It was just the tough membranes that wrap the brain, known as meninges. Even though Glenner and Wong had made sure that the vessels in these membranes were full of amyloid, brain amyloid did not have to be the same as blood vessel amyloid. The two scientists admit this in the discussion of their findings; they implicitly argue that the source of the amyloid they were looking at was the blood, not the brain. Nonetheless, since amyloid deposits in the blood vessels of the brain were only supposed to happen in Alzheimer's disease, they felt that they were on firm ground. Another place where they could have gone wrong is that they went through only a very simple series of steps to "purify" the amyloid. They relied almost entirely on the fact that amyloid is insoluble and can be stained with a special dye called Congo red. With the brilliance of hindsight, we know that the substance that they analyzed could well have been (and probably was to a certain extent) made up of an entire collection of proteins and other insoluble garbage, but fortune smiled.

The pair figured out the sequence of the first 24 amino acids of this brain vessel amyloid and published their findings in May 1984.[1] That was the good news. The disappointing news was that the sequence of those 24 amino acids didn't match a part of any protein that was known at the time. So, while the discovery was an important one and to this day serves as a true milestone in the history of Alzheimer's research, at the time it seemed to be only a partial solution. Undaunted, three months later, they reported on a second observation that expanded the importance of the original publication.[2] Researchers had recognized for some time that persons with Down syndrome developed dementia at a relatively young age, and when their brains were examined, they contained many of the same features that were found in the brain of Auguste D. Of these features, the one that interested the UCSD scientists the most was that the Down syndrome brain was loaded with amyloid. So Glenner and Wong repeated their experiments, this time using meninges from two males with Down syndrome. They found that the amyloid from the Down syndrome and Alzheimer's disease cases had almost exactly the same amino acid sequence—only one minor difference out of 24. They argued in their closing comments that the microscopic and biochemical identity of the two sources of amyloid suggested that Down syndrome was a "model" of Alzheimer's disease.

Of course, there are enormous differences in the symptoms of someone with Alzheimer's disease and someone with Down syndrome. So, what did Glenner and Wong mean in saying that Down syndrome was a model of Alzheimer's? The idea was that whatever havoc the amyloid deposits created in a human brain, the damage in Down syndrome should be similar to the damage in Alzheimer's disease. Other aspects of the two diseases could differ and differ dramatically. But there should be commonalities, and these could be hypothesized to be due to the presence of amyloid in the brain. Then they turned weakness into strength. We shouldn't worry that the amyloid they had characterized was from blood vessels, not from brain. Since amyloid was unique to Alzheimer's disease and Down syndrome, its presence in the blood vessels meant that it might leak into the blood. If it did, it could become the basis of a simple blood-based diagnostic test for Alzheimer's disease. No brain biopsy needed.

A SHORT INTERLUDE—SOME BASIC GENETICS

The next step in this story is a big one. To appreciate it fully, however, we are going to need a basic primer on modern molecular biology. We will start with what is known as the Central Dogma of Molecular Biology: DNA makes RNA makes protein. Without going into the chemistry of the process, the DNA part of this dogma represents our genes. The RNA part is a way of transferring the information from the blueprint of our genes in the cell's nucleus to cellular construction crews in the cell's cytoplasm. That's where a very big complicated collection of proteins (the ribosome) transforms the information in the RNA into a protein (like our aforementioned Important Brain Protein, or IBP). This two-step process allows the genes (our DNA) to remain somewhat protected from damage and also allows for multiple levels at which the process can be regulated. The real genius of the process, however, is the code that the cell uses to store the information and then transfer it from the gene to the protein.

Let's start backward and begin with the protein. All proteins are made of building blocks called amino acids strung together end to end to make a long chain. "Long" can mean sort of long; insulin, for example, is a string of 110 amino acids when it is first made. "Long" can also mean really, really long; the longest protein known is called titin. It's found in muscle and comes in at a staggering 26,926 amino acids. Even 110 amino acids could be really complicated; nearly 27,000 is biochemically terrifying. Fortunately, the cell simplifies things by using only 20 different amino acids. That means that no matter how long the string of amino acids in a protein is, it's just different ratios and orderings of the same 20 basic building blocks. The physics and chemistry cause the long string of amino acids to fold up in a very particular way. The result of this molecular origami is to form protein after protein with exactly the same complex shape, including parts that move and parts that stay rigid. These proteins are the way that our cells can do work.

The proteins are the end of the Central Dogma of Molecular Biology: DNA makes RNA makes protein. Our genes, we have learned, are the blueprints; they are the instructions that tell the cell how to build tens of thousands of different proteins. The way in which the information is stored in the blueprint before being translated into the finished protein product is a clever bit of bioengineering that takes advantage of the simplifying fact that you can build any protein in our body using the same 20 amino acids as long as you put them together in different combinations. Think of those 20 amino acids as letters and the protein as a word. Words, after all, are just strings of letters in a particular order that work as a group to convey a particular meaning. You can make thousands and thousands of words using only 26 letters (of course, this is not true for character-based languages such as Chinese). The same is true for proteins: you can make thousands and thousands of proteins using only 20 amino acids. It turns out that, just like our proteins, the DNA of our genes is also a long linear molecule. It too is made up of only a few building blocks, called nucleotides, and they too are strung end to end to make a long chain. That may sound direct and simple. Just store the information in the DNA with the same 20 "letters," and we're good. Sorry. DNA uses only four nucleotide "letters."

So how do we transfer the information in the linear four-letter nucleotide code of our genes to the 20-letter amino acid code of our proteins? With a lot of clever investigative work (recognized by multiple Nobel prizes), scientists figured out that DNA and RNA use not one but three nucleotides to encode a single "letter." That's called a triplet code, and it gives you enough encoding power to make 64 different "letters." Obviously, there is some redundancy, and indeed some amino acids are coded in six different ways by six different triplets. But let's not worry about that for now. The basics of the Central Dogma of Molecular Biology are this: three nucleotides of DNA tell the cell how to make three nucleotides of RNA which tell the ribosome which one of the 20 amino acids it should put in the protein. A rough diagram of this process can be found in figure 4.1.

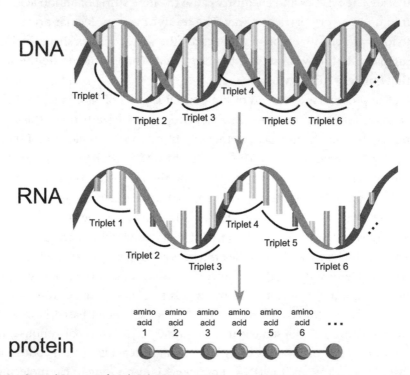

4.1 Central Dogma of Molecular Biology: DNA makes RNA makes protein.

Modern molecular biology has figured out this triplet code in its complete detail. That is fundamental to our story because one very powerful offshoot of knowing the code is that it has predictive power. It means that if you know the nucleotide sequence of a gene, you can very precisely predict the sequence of amino acids in the protein it codes for. The reverse is also true: if you have the sequence of amino acids in a protein, you can predict (with some error due to the redundancy) the sequence of nucleotides in a gene. Truly elegant.

RUNNING THE CENTRAL DOGMA OF MOLECULAR BIOLOGY BACKWARD

Now we are ready to go back to Glenner and Wong and discover why their publication was such a watershed moment in Alzheimer's disease research. By publishing a sequence of amino acids from the plaque protein, the amyloid beta (Aβ) peptide, Glenner and Wong basically put the keys to the kingdom on a table in the lobby of a busy train station. They were available for anyone to come pick them up and find the gene by translating the amino acid code of the Aβ plaque protein back into the genetic code and then go look in the human genome for where the gene for the plaque protein was.

Scientists are as ambitious as entrepreneurs and just as aggressive. It took just two-and-a-half years for three independent labs to find the gene that contained the code for the Aβ plaque protein. A fourth lab found the location of the gene, but not its sequence of nucleotides. Alois Alzheimer was no doubt smiling down on these announcements. Eighty years after his first full publication, the plaques he had seen in the brain of Auguste D. could be definitively linked to the chemistry and biology of the cell. And what a treasure trove of information this discovery opened. Perhaps most important of all, the linkage between Down syndrome and Alzheimer's disease, suspected for years, and predicted by Glenner and Wong on the basis of the sequence of the amyloid peptide, was proven beyond a shadow of a doubt. The gene for the plaque protein mapped to chromosome 21, the same chromosome for which there is one extra copy in Down syndrome.

Down syndrome is caused by a problem in counting chromosomes. Each of our chromosomes is represented twice in each of our cells. One comes from Mom; one comes from Dad. The dance of the chromosomes that is choreographed during the formation of an embryo is complex and orderly, but, like any process in real life, mistakes happen. If instead of one from each of your parents, one of your parents gives you two, you inherit a chromosomal nightmare. Now each cell in your body has an extra copy of that one chromosome. That extra copy means that all of the genes on that chromosome are not going to be regulated properly. Cells don't tolerate this situation very well. Too many chromosomes is usually a lethal situation. Sometimes, although there are serious problems, the result is not death. Down syndrome is one such example. It's the result of getting three copies of the second smallest chromosome, number 21. The gene for the Aβ peptide is found on chromosome 21, so persons with Down syndrome will have three copies (instead of two) of the Aβ gene. This means that all the cells of the body will make half again as much of the peptide as a normal cell would. The pathway to the buildup of plaques in the brain almost writes itself. Since the gene for the Aβ peptide is on chromosome 21, persons with Down syndrome make too much of it. Too much Aβ protein means too many plaques and too many plaques leads to Alzheimer's disease.

Another important gem in the treasure trove unlocked by this discovery was information about the nature of the protein that was the original source of the Aβ peptide. In biochemistry we call this larger protein a precursor protein. The most complete study of the plaque precursor was from the laboratory of Konrad Beyreuther and Benno Müller-Hill.[3] They were able to find the entire gene and thus were able to construct a map of the entire precursor protein. They discovered that the precursor was pretty long: 695 amino acids from one end to the other. That's no titin, but it's six-times-insulin long. Deeper analysis found that stretches of those 695 amino acids had familiar properties. One region in particular looked just like a part of a protein that would be comfortable sitting in a fatty environment. This was the region of the protein that was where the Aβ peptide was located. They guessed (correctly) that the protein was normally found sticking through the cell membrane and proposed that it was a cell surface receptor. A receptor is an important kind of protein that binds to something on the outside of the cell and tells the inside of the cell what it found. For example, there is a specific receptor that binds insulin on the outside of the cell and lets the inside know that there's insulin around. That lets the cell start gearing up to take up glucose.

Then things got even better. In the early 1990s geneticists studying the genes involved in familial Alzheimer's disease discovered that some families with this rare, inherited form of Alzheimer's had mutations that mapped very near to where the Aβ peptide gene is found on chromosome 21. Other labs reported similar findings from different families. Everything was coming together so nicely. All that was needed now was a name for the new protein. The field ultimately decided to call it the amyloid precursor protein—APP for short. There were a few other features of the APP protein that looked familiar, but there were no more clues as to what its normal function might be. We will come back to APP later when we go over the evidence for how it is involved in Alzheimer's disease. But right now, there are two more major discoveries from these heady 15 years that we need to look at.

PRYING THE AMYLOID PEPTIDE OUT OF ITS APP PARENT

The next discovery is actually an entire series of discoveries that led to the identification of the process by which Aβ is cut out of APP. With the discovery of the APP protein, the problem facing scientists is pretty simple to see. Look at the diagram of what APP looks like when it's in position in the cell membrane (see figure 4.2). The long chain of amino acids that makes up APP is indicated by the dark gray noodle. The small part of the APP protein that is the Aβ peptide is the light gray tube part in the middle of the dark gray noodle. The diagram is not to scale since Aβ is only 42 amino acids long while the entire APP protein is 695 amino acids, but for this part of the story the important stuff is all going on around the light gray tube. Here's the problem: the amyloid plaque is made up only by the small Aβ part that sits smack in the middle of the entire APP protein. If the Aβ peptide is going to get about the business of aggregating to make a deposit, it has to get rid of the rest of the APP protein. How does it get free so that it can make the plaques?

The answer to that question involves a type of protein known as a protease. That's a protein that's designed to cut other proteins into pieces. Think of proteases as molecular scissors that will only cut proteins. Proteases

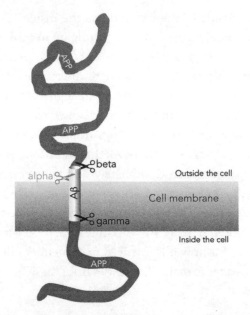

4.2 Amyloid beta (Aβ) peptide and its location in the amyloid precursor protein (APP) parent protein.

are usually pretty discriminating about which protein they will cut and where. Most often they cut only near a single type of amino acid (out of the 20 that are used). A protease would be exactly what we need to pry out the Aβ peptide from its APP precursor. One cut on each side of the light gray tube would do the job nicely. Researchers knew that these proteases existed long before they knew anything about them. As a side note, the particular proteases that cut out the Aβ peptide (for reasons that are lost in the mists of time) are known as secretases, but they are proteases all the same. They go by the Greek letters alpha (α)-, beta (β)-, and gamma (γ)-secretases.

Now let's get the Aβ peptide out of APP. The protease that starts the process is the β-secretase. Look again at figure 4.2. It's the black scissors to the right of Aβ on the outside of the cell membrane labeled "beta." It cuts APP right at the top end of the light gray tube. That leaves a whisker sticking out of the cell membrane and a long piece of dark gray noodle floating around in the space outside the cell. That piece of pasta probably has a function, and much energy has been spent looking for it, but we're going to stay focused on the Aβ peptide for now. The second protease, the one that finishes the job, is called the γ-secretase (the black scissors labeled "gamma"). Once its cut is finished, the Aβ peptide is free on both ends and can leave the membrane to go aggregate with other light gray tubes and make an amyloid plaque.

There is one more part to the story that I'll put in here for extra credit. It involves the third secretase, the one that we haven't mentioned yet, the α-secretase. If you look in the diagram at the light gray scissors to the left of the Aβ tube, you will see that they are poised to cut the APP protein closer to the membrane. That leaves less stubble sticking out of the membrane but, more importantly, it also cuts the Aβ peptide nearly in half. The γ-secretase still cuts in the same place, but the smaller part of the Aβ tube that is released by this second cut can't aggregate and form amyloid, and so it can't form plaques.

It's pretty clear that if Alzheimer and Kraepelin were right, then these three secretases are a huge part of the problem in Alzheimer's disease. APP by itself doesn't seem to be a problem, but if you cut it with the β- and γ-secretases, you start down the plaque-making road. We might guess that if you could find drugs that would block the ability of the secretases to cut APP, you would stop the plaques from forming and potentially stop Alzheimer's from happening. The α-secretase doesn't get anywhere near the same amount of attention, but we might also guess that if you could find a drug that would make it *more* active, you would destroy the light gray tube so that it couldn't make plaques, and that too might be a good thing. Less β- or γ-secretase activity or more α-secretase activity would seem to put us in a place where we would be less likely to come down with Alzheimer's disease.

How would you go about finding drugs like these? You could try to isolate the secretases. Then you could either set up an assay in a test tube where you could test compounds to find one that blocked the activity. Or if the protein is pure enough, you could do what Glenner and Wong did for APP and figure out the amino acid sequence of the secretase protein and then run the Central Dogma of Molecular Biology backward and go look for the gene. Many scientists were working these angles, but it turns out that this is a difficult problem. All three secretases are themselves located in the membrane, just like APP, and it turns out that membrane proteins are much more difficult to work with in test tubes than proteins that are normally found in the more watery environment inside the cell. The real breakthrough came once again from genetics, and the result was really pretty amazing.

Genetic studies had already uncovered a bunch of different mutations in the APP gene on chromosome 21 that led to early-onset familial forms of Alzheimer's disease. The genetics also told us that there were two more genes for familial Alzheimer's disease that remained unidentified. Scientists had used techniques of human genetics and mapped them to chromosome 1 and chromosome 14, but their map was very low resolution. There were dozens of genes in each region, and the scientists had no idea exactly where the Alzheimer's disease genes were in the DNA or what the protein was that they were responsible for. Then, in 1995 a team of geneticists in Toronto announced that they had done a high-resolution map of the area on chromosome 14 where they knew the Alzheimer's disease gene was located.[4] They reported that they had found a single gene with a pattern of mutation in five different families that identified it as the missing chromosome 14 Alzheimer's disease gene. They got the sequence of the gene, which told them exactly what the gene-encoded protein would look like. Shortly thereafter, based on the gene sequence of this chromosome 14 gene, a second very similar gene was found on chromosome 1 exactly where the genetics had said that the last familial Alzheimer's disease gene was located.[5] These two Alzheimer's disease genes became known as presenilin 1 (*PSEN1*; chromosome 14) and presenilin 2 (*PSEN2*; chromosome 1).

PSEN1 was the more common Alzheimer's disease gene. The predicted protein was smaller than APP, only 467 amino acids end to end, and just as with APP, there were portions of the protein that had familiar properties. Once again, there were stretches of amino acids that looked as though they would be very comfortable in a fatty cell membrane. Instead of one such region, however, they found seven (there are actually nine, but the group got pretty close). Like a thread in the hem of a fabric, they thought that the protein weaved in and out of membrane over and over. There was only one problem—there were no other clues, and so no one on the team had the foggiest idea what this protein might do. They speculated that it could be a docking site for other membrane proteins (like APP) or maybe a receptor or maybe a channel.

The next steps were agonizingly slow, but ultimately very satisfying from a scientific point of view. Many groups in many different countries worked to figure out what this protein was and what it did. Each group contributed more and more pieces to the puzzle. What became clear very quickly was that the presenilin proteins were associated with the γ-secretase. It also looked as though the mutant forms of presenilin cut APP slightly differently and the difference led to the formation of a more aggregation-prone form of the Aβ peptide. For our purposes, that is really all we need to know. The scientist in me loves the entire story, and it is tempting to tell the tale of the slow and winding path it took to prove that presenilin-1 and presenilin-2 were in fact proteases that lived in the cell membrane and were responsible for cutting the APP protein at the bottom end of our light gray tube, the Aβ peptide. But we need to stick to a more important story line.

Take a step back now, and think of the implications of the fact that the presenilins are actually the business ends of the γ-secretases. There were only three disease genes for Alzheimer's disease (even now, decades later, no new disease gene for Alzheimer's has been found). As a group, the three genes cause the rare familial forms of the disease, and their functions are closely related to each other. One is the APP gene itself, the source of the Aβ peptide; the other two are scissors (secretases) that help to free the Aβ peptide from its parent APP. There could be little doubt that Alois Alzheimer's instinct has been spot on. The amyloid plaques must cause the disease. Beyond mere correlation, the genetics clearly show that if you alter the precursor to the amyloid plaque peptide, or the fiddle with the proteases that help liberate it from its precursor, you cause aggressive, early-onset forms of Alzheimer's dementia. This has to be more than correlation; this must be causality.

Putting oneself back into this mindset makes it really easy to see why the symposium speaker I half-quoted in the beginning of this chapter was so full of confidence and optimism. All we needed were some anti-Aβ peptide drugs. Block the β- or γ-secretases, and we would be ready to move on to some other human disease with Alzheimer's disease nothing but a distant memory. Five years might have been a bit rushed even in these halcyon days of the field, but we were close, and we knew it.

But we were wrong.

WE'VE CURED THE MICE! WE ARE ALMOST THERE

We'll get to how and why we were wrong in later chapters. The full story of this wonderful 15 years would not be complete, however, without the telling of two more nearly miraculous findings. I've already told you that one way to avoid getting Alzheimer's disease is to not get old. There is actually a second surefire way to avoid Alzheimer's that I haven't told you yet: don't be human. It is a very curious and little understood fact that, except for us human beings, virtually no creature on planet Earth gets the plaque-infested form of dementia that Alzheimer described in Auguste D. A few species, including the mouse lemur, get close, but their amyloid plaques are less dense than those in humans. We still don't know why this is true, but it is. It's not for lack of trying on the parts of our fellow creatures. All animals age, and as they do their performance slips. We might not want to call it dementia, but it's a reasonable facsimile. More than that, almost all living creatures have an APP gene, and most have both β- and γ-secretase genes. So, the components are all there, but they just don't seem to come together. Enter genetic engineering.

Modern advances in molecular biology have made it possible to fiddle with the genes in just about any creature. We scientists use well-established tools that accomplish these feats all the time. Most biomedical research would in fact be impossible without them. The reason that biomedical research puts such a premium on their use is that they allow us to develop models of human diseases that speed the pace of the discovery process. This is particularly true for diseases where age is involved, such as Alzheimer's. An old human is 85–90 years old; an old mouse is 2–2.5 years old. A simple back-of-the-envelope calculation says, if you are trying to study the biology of an age-related disease, you can work 40 times faster using mice than using humans. To speed your work up even more, you might consider that an old fruit fly is six or seven weeks old, so you could work about 650 times faster using a fruit fly. Of course, you don't need me to point out that the brain of a fruit fly is not even remotely similar to ours; so, what's the point? That's a good question that I'll answer later. The mouse is far from perfect, but mice have most of the major brain regions that we do—only smaller—and over the years the genetic tools available in mice have become more and more powerful, making them a reasonable compromise between power, speed, and accuracy. But don't lose sight of what you already know without my telling you: a mouse is not a human. We can use mice as a first approximation, but mice don't naturally get anything resembling Alzheimer's disease. It's hard to study treatments aimed at fixing something if that something isn't there.

Remember this was a decade of unprecedented optimism and nearly religious faith in the power of genetics to solve any problem, no matter how complicated. The idea quickly emerged that perhaps if we could put the human APP gene into the genome of the mouse, then we would get a mouse with amyloid plaques and Alzheimer's disease. To be sure, this was a risky bet. If you look at the mouse APP gene and the human APP gene, they are almost 97 percent identical. APP does not seem to be the reason that mice don't get Alzheimer's disease. Of course, the Aβ peptide itself is a little less similar (only 93 percent identical), so there might be a modest difference in the tendency of the peptide to make amyloid. But good idea or not, what happened next was a furious competition to be the first lab to produce a strain of mice that carried the human APP gene with one of the mutations that caused early-onset familial Alzheimer's disease.

The first reported successes came in 1991, only four years after the identification of the APP gene. Three groups breathlessly told an expectant scientific world that they had done it. "Deposits of Amyloid Beta Protein in the Central Nervous System of Transgenic Mice" read the title of one of these announcements.[6] Unfortunately, this was not one of our field's finest hours. Within a year, two of these original papers had to be retracted, including the one with the title I just quoted.[7] It seems that one of the groups could not repeat their findings and the second group . . . well, let's just say that they thought that maybe they had messed up and that the picture of an amyloid plaque that they said they had seen in a mouse brain was actually a picture from a human brain. Fortunately, this embarrassing interlude was soon ended, and within a year multiple laboratories were reporting real success. When a human familial Alzheimer's disease mutant APP gene was introduced into the genome of mice, their brains developed real, honest-to-goodness Aβ plaques. The era of mouse models of Alzheimer's disease had begun.

So, what does a mouse with Alzheimer's disease look like? Well, just like people with Alzheimer's disease, there are a lot of plaques scattered all over the cerebral cortex, and the mice have trouble with spatial orientation, reminiscent of the wandering that is seen in humans. There is more to learn about these mouse models, but for now, think how gratifying it was to see these mice develop plaques as they got older—plaques that looked just like the human ones—simply by adding the human APP gene to the mouse genome. More validation of the tremendous enthusiasm in the field and the confidence that we were really close to a cure. And the fourth discovery of this period seemed to solidly seal the deal.

Perhaps the most important value of a good model is that it has predictive power. In the case of our mouse models, the logic would be that if we can get rid of the mouse plaques, we should be well on our way to a cure for Alzheimer's disease. This hope was very much on everyone's mind when a group at Elan Pharmaceuticals, headed by Dale Schenk, electrified the Alzheimer's community by reporting on an experiment that shouldn't have worked—a vaccine for amyloid.

There are four words in that scientific literature that, more than any others, hide a lot of misinformation and preconceived notions. Those words are "It is well-known that . . ." When you start a sentence with those words, you are telling your readers not to expect you, the author, to provide a link to any original source material to back up the statement that follows. Lots of times, that's a convenience that both reader and author benefit from, as in "It is well-known that Alzheimer's disease involves the accelerated loss of short-term memory." But those same four words can also inadvertently lock in a lot of misinformation. For example, in the early 1980s "it was well-known that" ulcers are caused by eating spicy food (they're not; most ulcers come from infections of the stomach with the bacterium known as *Helicobacter pylori*). In the late 1990s "it was well-known that" the Alzheimer's plaques were tough as concrete. You couldn't break them up and dissolve them even if you wanted to. Remember, this is how Glenner and Wong isolated the blood vessel plaques to begin with. Also, "it was well-known that" the brain was shielded from our immune systems. The shield is known as the blood-brain barrier, and it carefully regulates what gets in and out of our brains. Because of this blood-brain barrier, "it was well-known that" antibodies that our immune systems create in our blood to protect us from microbes do not protect our brain. That means that if you get a vaccine against a virus such as measles, the antibodies are not found in and among the cells of the brain. They fight of the virus in our blood and other tissues, but they are helpless if a virus gets into the brain. So, given that the plaques are like concrete and antibodies don't get into the brain, why in the world would you even try to develop a plaque vaccine as a way to prevent Alzheimer's?

I had always wanted to ask Dale Schenk what made him think that he should try this nutty experiment. I missed my chance because he was tragically taken from us by pancreatic cancer at far too young an age, but I remain curious to this day how the thinking behind this series of experiments evolved. What the Elan team did was to envision the plaques as an invading virus. Their idea was they would harness the power of the body's immune system to "teach" it to recognize and then eliminate the plaques in the same way that a flu vaccine "teaches" our immune system what influenza viruses look like so that it can recognize and eliminate them when real viruses come along. For the Alzheimer's vaccine, they just took some artificially synthesized Aβ peptide and began vaccinating their Alzheimer's disease model mice that they knew were going to develop plaques in their brains. The control mice just got saltwater. The experiment was a stunning success in no small part because "it was well-known that" it wouldn't work.

The results were some of the cleanest I have ever seen in all my years in science. The vaccine literally changed the fate of the Alzheimer's mice. The age-matched controls had a head full of plaques; the vaccinated mice had almost none. Equally remarkable was that the team could start the vaccination after the plaques had started to form. They showed that, over the time frame of their experiments, even preformed plaques appeared to recede in size and shape. I do not use the word electrifying lightly, but this Letter to *Nature* is truly one case where that adjective is earned. It is the fourth miraculous discovery of this wonderful era of Alzheimer's disease research. We had found the three known Alzheimer's disease genes; they were all part of the biochemistry needed to make the Aβ plaque peptide; we had engineered mice to carry a mutant human APP gene and had given mice plaques in their brain where before they had none; finally, we had developed a vaccine that cured the mice. The ways in which these findings could be turned into therapy in humans could not be clearer. With an unprecedented level of confidence, Elan started clinical trials almost immediately. Champagne corks were popping. Cigars were being lit. We were standing on the threshold of a cure for Alzheimer's disease.

That was 20 years ago.

In the intervening years we have crossed that threshold. Yet here, on the other side, we are no closer to treating or slowing Alzheimer's disease than we were. The unbridled optimism of those 15 wonderful years has faded to resignation or worse. The initial Elan clinical trials in humans had to be halted because of life-threatening side effects. More trials followed with different strategies for the vaccine, but what all of them have shown is that the mice were right, but we were wrong. We humans are indeed just like mice. If you give us a vaccine, we can clear the amyloid plaques out of our brains. But when we do, it doesn't help with our Alzheimer's disease.

This seemed like such a good way to study a human disease. What happened? Where did we go wrong?

WHAT HAPPENED TO OUR CURE?

Where did we go wrong?—a simple question with no simple answer.

We should not feel embarrassed at the idea of questioning medical researchers in this way; it is a question we are entitled to ask. Indeed, it is a question we must ask, over and over and over again, until we get an answer that is both understandable and satisfying. As a public, we bear the costs of this disease—both financial and emotional. We have invested heavily in its research—both clinical and basic. We have steadfastly supported this research with our tax dollars and with our bodies, as volunteers in clinical trials. We have been patient, but after all these years we have the right to ask, "Where is our cure?"

To start to answer this question, the next two chapters look at the thinking in the research community both during and after the 15 heady years at the end of the twentieth century. We will first take a critical look the strengths and weaknesses in the Alzheimer/Kraepelin model of the disease. This chapter will begin to unfurl all the red flags we rushed past as we rushed down the slope to what seemed like a sure gold medal at the end of our run. In chapter 5 we will meet for the first time what might be called the Central Dogma of Alzheimer's Disease—the amyloid cascade hypothesis. We will then go back and begin a methodical reexamination of all of the other theories—the ones that were pushed aside in the rush to embrace this attractive linear solution of the origins of Alzheimer's disease. We conclude this part with a chapter in which we turn back and reexamine the amyloid cascade hypothesis. We will have spoken only of its strengths and the strong rationale behind its original proposal. But we also need to go back and look at its many weaknesses. The pushed-aside theories are the scraps on the cutting-room floors of our research labs. They represent opportunities we missed that would have let us diversify our research portfolios. Many of these theories were perfectly compatible with the Central Dogma of Alzheimer's Disease, so why were they so thoroughly dismissed? What happened to them?

5

BUILDING A MODEL OF ALZHEIMER'S DISEASE

Let's pick up the threads of the story after the discovery of the APP gene, but before the identification of the presenilins or the story of the vaccine. Recall how miraculously simple it all seemed. As I said in chapter 4, the pathway almost writes itself: too much Aβ protein means too many plaques and too many plaques lead to Alzheimer's disease. We can represent that idea with a diagram that captures the basics (see figure 5.1). This is basically the Alzheimer/Kraepelin model with the biochemical source of the plaque added. From a researcher's point of view, however, this simple rendering is inadequate. The inadequacies are due to the fact that this simple model leaves out most of the underlying biological and biochemical details. That's a problem because it's in those details that the chemical wizards in the pharmaceutical industry will find the ways to synthesize the drugs we need to go after a cure. A key question that a biologist would want to ask is "How exactly do the plaques cause Alzheimer's disease?"

The gaps in the biological picture were apparent to most people in the field at the time, and attempts were made to fill in the missing details. The best known of these efforts was a 1992 publication by John Hardy and Gerald Higgins entitled "Alzheimer's Disease: The Amyloid Cascade Hypothesis."[1] It was a landmark publication that remains highly cited today. The details of the story have been modified over time, but the main tenants remain in place, as does the name itself—the amyloid

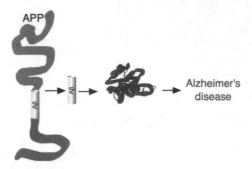

5.1   Early view of the causes of Alzheimer's disease. APP, amyloid precursor protein; Aβ, amyloid beta peptide.

cascade. The model that they put forward attempted to put the genetics and the neuropathology (the microscopic appearance of the brain) into register. It accomplished that aim and did so in a way that put the Aβ peptide at the center of the action. Looking back on this paper, 25 years after its initial publication, I find that the arguments have stood up well to the passage of time. The authors (both geneticists) had the following thesis as their central argument: "Deposition of the amyloid β protein . . . is the causative agent of Alzheimer's pathology and . . . the neurofibrillary tangles, cell loss, vascular damage and dementia follow as a direct result of this deposition."

Note that they did not say that the plaques cause Alzheimer's disease. They said the Aβ peptide caused it. The de-emphasis might have been either inadvertent or a compromise, but the distinction is an important one as we will come to see. They were also clear that while the facts at hand suggested that the Aβ peptide was the culprit, they did not have a clear idea of how the peptide did its work. They cite evidence from other laboratories that the Aβ peptide is toxic to neurons in a culture, but in the end, they leave it at that. Looking back, it's remarkable that the authors connected the dots as they did since neither the β- nor the γ-secretase was yet known. What was known at the time, indeed the instigation for the article, was that the APP protein was genetically linked to Alzheimer's disease. Yet rather than leave it at that, they explored biological processes that might link too much APP (Down syndrome) or the wrong kind of APP (the familial Alzheimer's disease mutations) to the massive loss of brain structure and function that comes with Alzheimer's disease. They presciently remarked about the possibility that APP, once inserted into the cell membrane, might still end up inside the cell. The way this happens is that a piece of membrane with APP stuck in it could be sucked back into the cell to make a vesicle (like blowing a bubble gum bubble backward). When that happens, the outside of the cell membrane is now the inside of the vesicle membrane. There was evidence that APP, following this pathway, ended up in lysosomes (remember those?), which, they speculated, might cut up the APP protein in ways that made the lysosome the ultimate source of Aβ.

They also considered the neurofibrillary tangles and the neuronal cell death that are so characteristic of various regions of the Alzheimer's disease brain and asked where they fit into the bigger picture. By this time, the tangle protein had been identified as a small protein called tau. Tau is found all over the cell (in fact, it's pretty much in all of the cells of our bodies), where it sits on long tubular structures called microtubules. Microtubules are like the beams and girders of the cell; they give it shape and tensile strength. When tau binds to the microtubules, it stiffens them. This stiffening or softening helps the cell not just keep it shape but also bend a bit so that it can twist or move around. Tau comes on and off the microtubule depending on whether or not it has been "decorated." Without going into the chemistry of it, the decoration of tau is a phosphate molecule. With a phosphate "decoration" tau doesn't bind microtubules so well. When it comes off, the microtubule girder softens just a bit. This is all well and good, but it seems that the proteins that add the decoration to tau (called kinases) get out of control in Alzheimer's disease. They are overactive, and as a result they add too many phosphates. Too much tau comes off the microtubules, and the microtubules go all limp. Equally important, however, when tau is overphosphorylated, it starts to aggregate. It doesn't make a β-pleated sheet, so there is no tau-based amyloid, but it aggregates, nonetheless. The end result of this tau aggregation process is the threadlike tangle that Alzheimer saw as a "ghost" where a nerve cell had once been.

What intrigued Hardy and Higgins about the plaques and tangles situation was that the Aβ peptide was believed to regulate calcium, although again no one knew how. Also, no one knew yet what the identity of the tau kinase (the phosphate decorator) might be. What was known at the time that the two scientists were writing their review was that the activity levels of some kinases were regulated by calcium. This led the authors to suggest that by regulating calcium and hence the tau kinases, the Aβ peptide could also be responsible for the tangles. Their idea was that the plaques and tangles, rather than being two events caused independently by the Alzheimer's disease process, were actually stepwise events in a single line of causality. The Aβ peptide, through its regulation of calcium, caused tau to have too many phosphate decorations, and that allowed the tangles to happen.

The calcium angle was really a twofer for them. That's because it had been repeatedly shown that when the levels of calcium inside of a nerve cell get to be too high, this triggers cell death (actually a cell suicide). If it were true that the Aβ peptide could increase the levels of calcium inside the cell, then it might be responsible both for the tangles and for the neuron cell death.

And to bring it on home, neuron cell death will disrupt network connections in the brain. Those disruptions will lead to more and more malfunctioning circuits—like the memory circuit. This, then, would explain the dementia. They conclude their paper with a tip of their hats to the labs that were developing mice carrying human APP genes in their genome and a hope that researchers would use these animals to develop a cure for Alzheimer's disease.

We can diagram Hardy's and Higgin's ideas using the same graphic elements we used in the first figure (see figure 5.2). Doing so reveals a much richer biological description of Alzheimer's and Kraepelin's original ideas, although one wonders why the authors thought of this as a cascade. There are many downstream consequences to the generation of Aβ from the APP precursor, and when I look at the diagram I find that the bad things that happen in Alzheimer's disease don't "cascade" so much as "spray" off the Aβ peptide in almost every direction. Small quibbles aside, this was a great paper for its time. It gave prominence to all the pathologies, not just the plaques. It incorporated the tangles, the blood vessel amyloid, and the nerve cell death into the mix of problems. Before the genetics had pointed in this direction, it gave full voice to the possible role of cellular vesicles, including the lysosomes, in the generation of the Aβ peptide. And it brought up calcium dysregulation as a potential part of the disease mechanism. Each of these features has stood up well to the passage of time. And again, these ideas were floated in the early 1990s when we had much less information than we do today. Presented as a

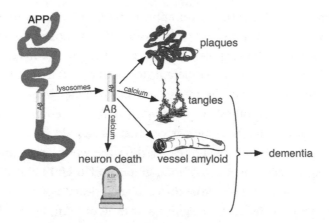

5.2 A diagram of the amyloid cascade hypothesis adapted from its original conception by Hardy and Higgins. APP, amyloid precursor protein; Aβ, amyloid beta peptide.
Source: J. A. Hardy and G. A. Higgins, "Alzheimer's Disease: The Amyloid Cascade Hypothesis," *Science* 256 (1992): 184–185.

hypothesis, it was a timely extension of the philosophy of Alzheimer and Kraepelin that brain structure drives brain function, and when the structure goes bad, so does the function. In extending these ideas, it married them to the genetic data linking Alzheimer's disease to APP mutations and Down syndrome. This allowed the authors to make their hypothesis richer in biological and chemical details and offer potential paths toward therapies aimed at blocking the development of Alzheimer's disease.

But it's 25 years later, and we still don't have a cure. What happened? What happened was that reality intervened. The value of a truly good hypothesis is that it can be tested to establish its strengths or weaknesses. Testing a hypothesis is hard work, in no small part because it's nearly impossible to prove that any hypothesis is 100 percent correct. Instead, most hypothesis testing is about looking at a variety of different circumstances to find out when it is wrong. This is how science works, and it is a highly effective way to advance a field. In science, as in life, most seemingly black-or-white questions have "gray" as the correct answer. Learning where and when a hypothesis is wrong establishes boundaries to the original idea and is actually quite useful.

The idea that it is almost impossible to prove that a hypothesis is 100 percent correct is an important concept that is often confusing to nonscientists. Let's use the worm/apple hypothesis again as an example. We tested our hypothesis that worms cause the skin of an apple to turn red by looking at 10 red apples. According to that test, our hypothesis was correct. But we didn't go to a different store and test 10 red apples stocked by a different grocer. Nor did we test any green apples. When we did, we proved that our hypothesis was wrong in both of these new situations. We agreed (at least I hope we agreed) that we should largely reject the original hypothesis. Yet even after rejecting the worm/apple hypothesis, we had nonetheless acquired a good amount of useful data. We clearly learned never to buy apples from the cheap and sleazy grocer who was selling the ones collected from the ground, instead of from the tree. That put us in a position to reformulate the hypothesis as a worm/grocer hypothesis instead of a worm/apple hypothesis. This "tweak" is significant, but it builds on our first investigation, is consistent with all of the data of our "experiments," and is much more robust in its predictions than our original hypothesis.

The idea that a good hypothesis can be wrong yet still be valuable brings us back to the Hardy and Higgins amyloid cascade hypothesis. Almost immediately after its publication, researchers began testing the amyloid cascade hypothesis and observations began accumulating that its predictions were sometimes wrong. This should not have been a problem since there were tweaks that could have adjusted its predictions to fit the new data. But something had happened in the field. The momentum of the 15 golden years had gathered so much force that the ideas in the amyloid cascade hypothesis were nearly unstoppable—so much so that more and more people stopped listening to their own data. And that's a terrible thing to do to a perfectly good hypothesis.

In 1992, the year that the Hardy and Higgins paper was published, I had just begun my own studies of Alzheimer's disease after studying the genetic control of early brain development for many years. I was eager to apply the lessons of development to the problems of aging and to see how they related to Alzheimer's disease. I was open to any good ideas. Shortly after I began my studies in my new home in the Alzheimer's Research Laboratory of Case Western Reserve University, our group's External Advisory Board paid a scheduled visit to hear about our progress and counsel us on our future directions. When the advisors sat us down for our presentations, I was pretty excited because I had begun to build a connection between the nerve cell death in Alzheimer's and a paradoxical loss of cell cycle control in nerve cells. For reasons that we still don't fully understand, neurons stay in a permanently nondividing state. I proudly announced my hypothesis that in Alzheimer's disease neurons are tricked into reentering the cell cycle and, since they can't divide, they die. I showed my data and added that it was almost as if Alzheimer's was a type of cancer—a loss of cell cycle regulation.

Our advisors were quiet as they digested these ideas. Finally, the chair of the committee spoke, and I still distinctly remember that I was given a strict warning. "Son, if you're not studying amyloid, you're not studying Alzheimer's." I was crestfallen that my grand new hypothesis had met with such a tepid response. But more than the rejection of my new idea, I was struck by the solemnity of the warning and the certainty that I had to stick to studies of amyloid. Why, I wondered? Even though I was new to the field, it was already apparent to me that there were lots of other interesting questions to look at. Plus, even if it was the major player, there would still be important details apart from amyloid itself that would influence the course of the disease. Why was amyloid the only piece of the Alzheimer's puzzle that was worth working with?

THE AMYLOID CASCADE HYPOTHESIS
GROWS . . . INTO A BULLY

Looking back with a 25- to 30-year perspective on that question, I find that a big part of the answer is hidden in a second paper that was published on the tenth anniversary of the Hardy and Higgins paper.[2] It was another short opinion piece that appeared in *Science*, the same magazine as the original amyloid cascade hypothesis piece. Hardy was also an author on this work, but this time he was joined not by Gerald Higgins but by Dennis Selkoe. The science of the field had advanced by leaps and bounds. All three of the secretases were now known, and chemical inhibitors were being developed for the β- and γ-secretases. The idea that the Aβ peptide was the cause of the disease was very much in play, but the differences between the two papers serves, I believe, as a revealing way to track the evolution of the thinking in the field.

The pathway to dementia described in the 2002 paper was expanded, and the diagram of the amyloid cascade hypothesis made the idea of a "cascade" much more visible (see figure 5.3). The starting point was still too much Aβ peptide, but the authors gave great prominence to the recent discoveries that there were stages of aggregation that could be identified as having different levels of toxicity. They came down in favor of the idea that it was the intermediate steps of the aggregation process that were the truly evil ones. Aβ peptides, floating around solo, were not a problem; a plaque full of thousands of them was not a problem; it was the intermediate phase where only a few Aβ peptides were stuck together that was the real problem. These halfway aggregates were termed oligomers. The authors also pointed out that most of the familial Alzheimer's disease mutations that had been found so far—in either APP or one of the presenilins—led to a slightly longer Aβ peptide (42 amino acids instead of 40) that was a bit stickier and more prone to form aggregates. Tau was dismissed because of the fact that mutations in the tau gene were correlated with a different type of dementia, not Alzheimer's disease. Hardy and Selkoe also included inflammation in the steps of the new cascade and made reference to oxidative damage. But these features of Alzheimer's disease chemistry were not sketched out in any detail. In the end, they diagrammed their ideas with a text-only figure. I've tried to reproduce their image using the symbols of the earlier figures (see figure 5.3).

The authors acknowledge that there were soft spots in the hypothesis. Tellingly, the subtitle of the paper included the words "Progress and Problems." Nonetheless, they were full of optimism that a cure was at hand based on the amyloid cascade hypothesis: "The development of anti-Aβ therapeutics remains a rational approach to treating AD [Alzheimer's disease], based on our current understanding of the earliest features of this disease."

I've chosen this paper as an example of the thinking in the field for three reasons. The first is that it is nearly identical to the original paper in its format—a miniature review in the same high-profile journal—published explicitly as a 10-year retrospective on the original amyloid cascade hypothesis. The second reason is that it captures the evolution of the thinking in the field not because of what is there, but because of what is missing. The Hardy and Higgins acknowledgment that lysosomes could be involved in the creation of the Aβ peptide is gone. The role of calcium

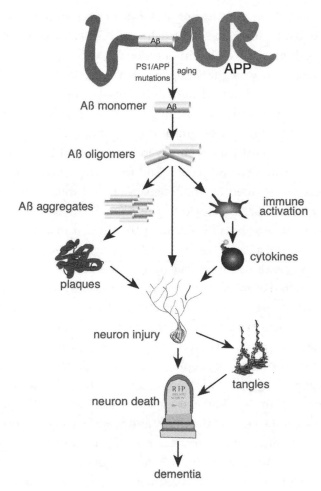

5.3 A diagram adapted from the revised amyloid cascade hypothesis as envisioned by Hardy and Selkoe in 2002. APP, amyloid precursor protein; PS1, presenilin-1; Aβ, amyloid beta peptide.
Source: J. Hardy and D. J. Selkoe, "The Amyloid Hypothesis of Alzheimer's Disease: Progress and Problems on the Road to Therapeutics," *Science* 297 (2002): 353–356.

has vanished. The amyloid of the blood vessels has also disappeared from the story. No new data had been found in the 10 years between 1992 and 2002 that would justify any of these exclusions.

The third reason that I've chosen this paper as an example is that the structure of the paper says a lot about the struggles that were being waged over the field's attempt to test the amyloid cascade hypothesis and where those tests stood 10 years after its original proposal. The paper is four pages

long, just like Hardy and Higgins. But instead of expanding a single logical case for a unifying model of Alzheimer's disease, the Hardy and Selkoe
article has three parts, each with a different purpose. The first is a one-
column reprise of why the stripped-down amyloid cascade hypothesis
(minus lysosomes, minus calcium, minus vessel amyloid) is still a strong
candidate for describing how Alzheimer's disease happens. The second
section is three times the length of the first and deals with rebuttals to
six self-identified criticisms of the hypothesis. The third part is about the
same length as the second section and represents musings on how the
hypothesis can be used in the design of clinical treatments, all except
one of which is based on amyloid. This paper accurately captures a vein
of thought that was broadly shared at the time. It marks how the field
transitioned from a broad-based view of the origins of Alzheimer's disease
to a simpler one-cause-fits-all imagining of the disease.

The attraction of this newer rendering of the amyloid cascade hypothesis was due in large part to the genetic findings surrounding the causes
of familial Alzheimer's disease and the connection to Down syndrome. It
just seemed that piece after piece of the puzzle was falling into place. As
a result, a good number of people in the field were convinced that it was
going to prevail. Everyone was optimistic. Some became nearly strident in
their defense of the amyloid cascade as the sole cause of Alzheimer's disease. The second section of the Hardy and Selkoe paper captures that feeling. Objection after objection is acknowledged but dismissed. The logic
used in the argument for dismissal is not in question. For any one objection to amyloid as a basis of Alzheimer's disease, the authors' response
in the paper is a reasonable counterargument. By the end of the middle
section of the paper, however, the quantitative impression is overwhelm-
ing. There are simply too many objections. Thus, however logical each
individual response might be, there is a message in the numbers. By rais-
ing six objections from the field (and there were no doubt others), they
tacitly admit that the hypothesis is under serious attack, as indeed it was.

I would argue that the appropriate response to this admission would
have been different than the one the authors chose. It would have been
much better if they had not minimized the widespread objections, acting
as if the matter were closed ("In summary, none of the currently perceived

weaknesses of the amyloid hypothesis provides a compelling reason to abandon this idea").

The literature at the time had opinion pieces challenging the amyloid cascade hypothesis, and an appropriate scientific response would have been to suggest modifications that could be used to bring its predictions more in line with the ever-increasing body of data. Dismissal, as attempted by so many of the amyloid proponents, is a debating tactic more than a scientific argument. The adherents of the amyloid cascade hypothesis had an obligation to support as full an exploration of the biology as possible. Rather than take that approach, however, the defense of the hypothesis became the mission, in and of itself. This was where the field was at. Hardy and Selkoe, our spokesmen of this approach, do acknowledge that "together [the objections] certainly point to important gaps in our understanding of AD." But re-creating the sentiment of the time, I find that these words were just ways of letting them hedge their bets. Remember, at the same time, the learned men of our External Advisory Board admonished us, "If you're not studying amyloid, you're not studying Alzheimer's."

OTHER OPTIONS: INFLAMMATION OF THE BRAIN AS A CAUSE OF ALZHEIMER'S DISEASE

In suggesting that we only look at amyloid, our advisors were debating, not advising, and the 2002 Hardy and Selkoe paper is symptomatic of that attitude. There were plenty of other hypotheses out there, all of them had merit, and from the perspective of anyone who might look in from outside the field, the debate they were trying to have was a silly one. That's because accepting one of these other theories as valid did not require that the amyloid cascade be abandoned as a contributor to Alzheimer's disease. It just needed to be tweaked. Very few of the other ideas required anyone to completely reject amyloid as part of its thesis. In fact, almost all of them worked well right alongside the amyloid cascade hypothesis. This was truer for some than for others, but overall there was no good reason to sweep the entire collection of viable alternative hypotheses under the rug or insist, as did our advisors, that the only path to Alzheimer's disease was through amyloid. We've spoken about almost all of these alternatives in chapter 3. Now we need to revisit them and ask about their relationship with the amyloid cascade hypothesis.

As we have learned, there are strong genetic, epidemiological, and microscopic (neuropathological) pointers to the role of brain inflammation in the Alzheimer's disease process. Some of the strongest proponents of this idea were Patrick and Edith McGeer, Joe Rogers, Sue Griffith, and others. In the mid-1990s, these authors argued that there was substantial evidence that inflammation was deeply involved in the Alzheimer's disease story.[3] The evidence could be found in the epidemiology of people with rheumatoid arthritis who received aggressive anti-inflammatory treatment and also in the appearance of cellular and biochemical signs of chronic inflammation in the Alzheimer's disease brain. Based on these data, the researchers hypothesized that the real driver of Alzheimer's disease was the toxic consequences of an unrelenting inflammatory process. They drew attention to the fact that the plaques and tangles themselves were not just pure Aβ peptide or pure tau. Rather, they contained as many as 40 other types of proteins, most of which were linked in one way or another to an active immune system response. The inflammation hypothesis makes many clear predictions about therapeutic approaches to Alzheimer's disease as did the amyloid cascade hypothesis. Unfortunately, there have been only lackluster and half-hearted efforts made toward testing them in humans, especially compared to the effort and dollars that have been invested in testing the amyloid cascade hypothesis.

You might imagine that the debate between the inflammation proponents and the amyloid supporters was a battle royal between two completely incompatible views of Alzheimer's disease—a fight to the death where the success of one hypothesis required the complete suppression of the other. That was certainly the view of the people pushing the amyloid cascade hypothesis. Inflammation was just a distraction from their opinion. But in fact, the inflammation hypothesis folks never argued that plaques or tangles should be excluded from the story of Alzheimer's disease. Far from it. In fact, they argued that one very likely trigger for the inflammatory reaction was the plaques and the tangles themselves. This meant that the inflammation-based model was actually a combined model. Its proponents emphasized the role of inflammation but fully incorporated the steps of the amyloid cascade hypothesis—a big-tent approach. To be sure, there was a difference in emphasis. The proposition was that "much of the neuronal damage observed in the disease is caused not by the fundamental pathology itself but by the inflammatory response to it."[4]

No shrinking violets, the inflammation advocates pointed out a number of areas where the amyloid cascade hypothesis might come up short, and some of these objections found their way into the Hardy and Selkoe rebuttal section. But rereading the pro-inflammation arguments from two decades ago, I find nothing exclusionary about their arguments. There was no suggestion that the idea of amyloid playing a role in the Alzheimer's disease process needed to be abandoned. Yet the amyloid proponents offered little of the big-tent philosophy in return. Even in arguing for testing an anti-inflammatory approach (a two-sentence suggestion), Hardy and Selkoe are careful to point out that the drugs may actually be working through blocking APP cleavage by the γ-secretase (the second of the two sentences). Basically, the amyloid field was saying to the inflammation field, "If you aren't studying amyloid, you aren't studying Alzheimer's."

OTHER OPTIONS: POOR FAT MANAGEMENT AS A CAUSE OF ALZHEIMER'S DISEASE

The idea that lipids, in particular cholesterol, might be playing an important role in setting the risk for Alzheimer's disease is an attractive one. As early as the 1960s, researchers had discovered that there were reduced levels of different kinds of lipids, including cholesterol, in the brains of persons with Alzheimer's disease.[5] What really gave this concept a push, however, was the discovery of the enormous impact of the *APOE* gene on the risk for Alzheimer's disease. Remember that the APOE protein functions as a fat (lipid) shuttle, for cholesterol in particular. It allows lipids to be moved from one place to another in the watery environment of our body. Having the "wrong" kind of APOE protein in your body can increase your risk of Alzheimer's disease by 10-fold. The paper announcing its dramatic effects was published in 1993,[6] just one year after the Hardy and Higgins paper.

Researchers quickly focused on the cholesterol part of that story. That was in part because of reports in the cardiovascular field that defects in

APOE were responsible for increasing the risk of coronary heart disease. One thought was that the increased risk of coronary disease was due to the fact that APOE4 (the high-risk form of APOE) was pretty bad at carrying cholesterol. That might mean that there was too much cholesterol in the blood. Epidemiologists began to sift through the cognitive data of persons who had been seeing heart doctors for high cholesterol and had been taking drugs that block the cell's ability to make cholesterol. As a group, these drugs are known as statins. As with the NSAID/inflammation data, the epidemiologists hit pay dirt again. Reports began to appear around 2000 that persons who had taken statins had a significantly lower risk of Alzheimer's disease.[7]

This all seems like good news, but the biologists were stumped. The ideas around fat management never really hung together into a formal theory of fats-cause-Alzheimer's disease because it wasn't really clear to the lipid people how these fat-handling problems, even focusing on cholesterol, were leading to the symptoms of Alzheimer's disease. Lipids are hard to work with, whether you're a cell or a scientist. So, testing any idea in a laboratory setting is tricky. There was evidence that certain types of lipids were associated with a premature death of certain brain cells and that the "ping-squirt-ping" process at the synapse between two neurons was also lipid sensitive. But this would apply to every neuron in every part of the brain and at every age. It didn't really give a clear insight into where the particular condition we call Alzheimer's disease comes from.

Did these thoughts and musings exclude the ideas of the amyloid cascade hypothesis? Though they were struggling with what the biochemistry and the epidemiology were trying to tell them, the lipid chemists, like the inflammation proponents, were only too happy to share their model building with the amyloid theorists. They offered up several ways in which the amyloid story and the lipid story fit together. They pointed out that changing the lipid composition of the cell membrane would change the properties of any protein found there. That included APP and all of its secretases because they all have homes in the membrane. There was even some evidence presented that the lipids altered the tendency of the Aβ peptide to aggregate in the first place. The biggest olive branch of all, however, was that the lipid people pointed out that not only could APOE bind lipids and cholesterol but it could also bind the Aβ peptide. You would have thought that this would have increased the interest in lipids among the amyloid folks. You would have been wrong. What the APOE finding really accomplished was to induce an entire raft of speculations as to how the elevated risk of Alzheimer's disease that came with the APOE4 gene probably had nothing to do with lipids. In other words, rather than share the spotlight, the amyloid proponents dismissed any role for lipids at all. The amyloid tent was getting smaller and smaller. This suppression of other ideas reached the point that even when statins were tested for their efficacy in Alzheimer's disease trials, the primary outcome measure the investigators were looking for was a change in the level of Aβ in the cerebrospinal fluid. Cognition and other measures were secondary. In this hall of mirrors, primary outcomes were never met, and today, there are virtually no ongoing trials of agents that alter lipids (including cholesterol) or modify their function.[8]

OTHER OPTIONS: LOSS OF MYELIN INTEGRITY AS A CAUSE OF ALZHEIMER'S DISEASE

One person who took more than a passing interest in the linkage between fat management and Alzheimer's disease was a UCLA psychiatrist named George Bartzokis. Bartzokis worked with MRI images of living human brains. He cared about nerve cells, of course, but mostly he cared about myelin. He would tweak the magnets and the detectors on his MRI machine so that he could see the myelin of the brain more clearly (T2 weighting). What he found, after looking at lots of different subjects at different ages and different levels of cognitive abilities, convinced him that the loss of myelin actually lay at the root of the cognitive problems of Alzheimer's disease. He discovered that the amount of myelin increases until we are middle-aged and then begins to decline. He stressed how closely this correlated with the cognitive capabilities of the brain regions he was looking at. Then he looked at persons who had been diagnosed with Alzheimer's disease. What he found was pretty exciting: they had significantly less myelin in their brain than would have been expected from the data he had accumulated from persons without dementia.

INTERLUDE: AN OPTIONAL DEEP DIVE INTO THE REALITIES OF HUMAN SCIENTIFIC DATA

It is worth taking a look at the actual data from the Bartzokis paper.[9] The reason is that it will give you a flavor of the kind of data that Alzheimer's disease researchers have to deal with. Human beings are absolutely terrible as research subjects. One look at the "raw" data from a typical human experiment will show you why. In figure 5.4, each symbol represents one person. The further the symbol is toward the right of the graph, the older that person was when they went in to get their brain imaged in an MRI. The further the symbol is toward the top of the graph, the more myelin they had in their brain (frontal cortex to be precise). Each circle represents a person who was normal in their cognition (no dementia) when they were imaged. The curved line going through the middle of the points is the "best fit" to this control data. Each black square represents a single person who had a clinical diagnosis of Alzheimer's disease when they went into the magnet. The authors came up with two hypotheses based on this data. First, myelin content goes up until middle age and then goes down. Second, people with Alzheimer's disease lose myelin much faster than people with no dementia. The authors performed a statistical analysis of the data to support their hypothesis.

But look at the actual data that power these conclusions. I would suggest that the "messy" nature of their story offers some insights that should help you read any article about human health research. The most important insight is how much variability there is from person to person. Look just at the circles; they look more like a cloud than a curved line. You can see how, at any age, some people have a lot of myelin and some people have much less. Looking at the black squares, you can also see that some people with Alzheimer's disease have more myelin than some of the controls at the same age, and some people with no dementia have less myelin than some of the people with Alzheimer's disease. Look at the dashed line. We can trust the authors that it is a mathematically correct description of the controls. Be that as it may, for any one person whose point falls above the curve (more myelin than you would expect for someone at that age), it does not mean they're intellectually gifted. For a person below the curve (less myelin than expected at that age), your poor myelin showing does not mean that you are suffering from

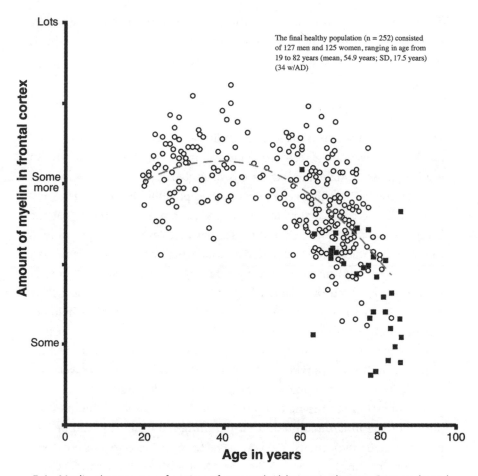

The final healthy population (n = 252) consisted of 127 men and 125 women, ranging in age from 19 to 82 years (mean, 54.9 years; SD, 17.5 years) (34 w/AD)

5.4 Myelin changes as a function of age and Alzheimer's disease. Data replotted from source.
Source: G. Bartzokis, J. L. Cummings, D. Sultzer, V. W. Henderson, K. H. Nuechterlein, and J. Mintz, "White Matter Structural Integrity in Healthy Aging Adults and Patients with Alzheimer Disease: A Magnetic Resonance Imaging Study," *Archives of Neurology* 60 (2003): 393–398.

dementia. It's a bit confusing, and the huge variability requires especially rigorous statistical analysis before any conclusions can be drawn. This is what real human research data look like—warts and all.

BACK TO MYELIN

The correlation between the loss of myelin health and the onset of Alzheimer's disease was and is really compelling. First of all, the idea that myelin is a big player in Alzheimer's disease is strongly supported by connections between the fat-handling gene network (and possibly the vesicle gene network) and the construction of myelin in the brain. Remember that the myelin membrane is fat rich and loaded with cholesterol. Also, as early as the late 1960s researchers had started to notice that myelin loss was a big part of the loss of brain mass that was found in Alzheimer's disease. It doesn't seem like too much of a stretch in this case to imagine that these links among the microscopy, the biochemistry, and the genetics of Alzheimer's disease might be trying to tell us that there is an important causal relationship between Alzheimer's and the processes that keep myelin healthy. Thinking in this way provides lots of clues for putting together new disease models, not to mention for developing new types of drugs.

The question for the field, then, would be the same as the question raised for the inflammation and fat-handling hypothesis: If you believe in the myelin hypothesis, do you have to stop believing in amyloid? The answer from the myelin folks was, "Not entirely, but sort of." Or, as Bartzokis himself put it, "The myelin model does not discount contribution of toxic species such as Aβ (as well as tau . . .) but rather proposes overarching 'upstream' mechanisms that trigger both Aβ and tau . . ."[10]

In the language of chapter 2, he's arguing that this is a three-body problem with myelin being the x-factor (the star in figure 2.1) that causes both amyloid and dementia. He adds to his argument by pointing out that in addition to their role in the production of the Aβ peptide, the β- and γ-secretases are also used by the oligodendrocyte to regulate myelin thickness. In doing this, he is arguing that the presenilin genes (γ-secretase) are actually Alzheimer's disease genes not because of their role in Aβ formation but because of their role in myelin. If true, this seriously undercuts one of the strongest legs of support for the amyloid cascade hypothesis.

Looking back over the history of the myelin hypothesis of Alzheimer's disease, I see it as aggressive but inclusive. It is aggressive in that it seeks dominance over amyloid. The idea is that the myelin changes are upstream of those found in amyloid and tau. That necessarily diminishes the importance of amyloid, but it doesn't exclude it. It recognizes the existence of amyloid. It acknowledges that amyloid might contribute to the deterioration of Alzheimer's. More importantly, the theory works to incorporate the major ideas of the amyloid cascade hypothesis into the model, thus making it richer and more biological.

So, what was the response of the amyloid partisans? Crickets. The 2004 review by Bartzokis laying out the myelin hypothesis[11] was cited by other people just over 300 times during the 10 years after its publication. That's respectable, but the Hardy and Selkoe paper, for comparison, was cited over 3,200 times during the same 10-year period after it appeared. This is a rough measure of how thoroughly nonamyloid ideas were kept out of the field. The myelin ideas were never seriously tested. They were "debated" into nonexistence and never granted a seat at the table where the mechanisms of Alzheimer's disease were being worked out. Today, myelin is not even cited as a secondary outcome measure worth taking.

OTHER OPTIONS: IMPROPER VESICLE MANAGEMENT AS A CAUSE OF ALZHEIMER'S DISEASE

Before the evidence from the genetics of late-onset Alzheimer's disease pointed squarely to the relationship to genes for vesicle proteins, there was already considerable evidence that improper vesicle management was a factor to be understood. Lysosomes in particular had been singled out for study. As early as 1967, there was evidence that the amyloid plaques of the Alzheimer's brain contained proteins (proteases) that should normally be found in only lysosomes. Following up this evidence, scientists found that vesicle abnormalities were some of the earliest to develop in Alzheimer's disease. The connection to the vesicular system and the process of autophagy remains an active area of research to this day. The presenilin proteins, for example, help to keep the inside of the lysosomes much more acidic than the rest of the cell. By doing so, they make sure that the proteases inside the lysosome work at peak efficiency.

This provides a link to the early-onset familial forms of Alzheimer's disease, as well as to the more common late-onset form.

As with inflammation and myelin and fats, the proponents of the vesicle hypothesis of Alzheimer's disease fully incorporated amyloid into their models of the disease. They repeatedly pointed out that the lysosome problems could easily account for a buildup in the Aβ peptide since it was most likely created in lysosomes to begin with. Recall that this was a point raised by Hardy and Higgins in their original paper. The attitude in the vesicle camp was pretty similar to that in the myelin camp—aggressive but inclusive. They had worked hard to show how vesicles and lysosomes could be the core of the Alzheimer's disease problem. Yet at the same time, they recognized that amyloid came with Alzheimer's and were open to including it in their models.

Again, we ask what was the response from the proponents of the amyloid cascade hypothesis to the notion that the lysosomes were a part of the story of Alzheimer's disease? Unfortunately, the answer to the question is also the same. There was no response. In fact, once again, rather than open hostility there was a studied silence that amounted to an active avoidance. Remember that the first paper on the amyloid cascade hypothesis, the Hardy and Higgins paper from 1992, prominently referenced the lysosome as the probable cellular source of the Aβ peptide. The 2002 paper by Hardy and Selkoe had no reference to lysosomes at all. Literally, the word lysosome does not appear anywhere in the four-page 2002 paper. It seems that the amyloid folks at the External Advisory Board were sticking to their guns: "If you aren't studying amyloid, you aren't studying Alzheimer's."

OTHER OPTIONS: OXIDATIVE DAMAGE AND MITOCHONDRIAL DYSFUNCTION AS A CAUSE OF ALZHEIMER'S DISEASE

We haven't touched much on the topic of oxidation yet. We need to correct that omission because there is a long history of aging research that views oxidation as a critical component of the aging process. Simply speaking, oxidation is rust. When we're talking about iron or steel, we all recognize it when we see it. And if we see too much of it on the girders of a bridge, we think long and hard before we drive our car over it. To a chemist, what we call rust is the addition of an atom or two of oxygen to an atom of iron. When we're talking about biological oxidation, we have to get deep into the chemistry of electron addition or subtraction (which we will not). In a cell, everything rusts eventually. Stated in a more formal way, biological oxidation can happen to virtually any molecule in a cell—a lipid in the membrane, a sugar in the cytoplasm, a nucleotide of the DNA in the nucleus—anything. You don't need to be a chemist to understand what rust does to a steel beam. Now imagine that same process at work on a cellular structure. Cells with oxidized parts don't collapse in the same way a bridge with rusted beams collapses, but the biological consequences of cellular oxidation are just as insidious.

With the rusted bridge analogy in mind, you can understand why, as research turned up more and more evidence of oxidative damage during the course of Alzheimer's disease, there was a strong push to develop this linkage into a theory of Alzheimer's. People like George Perry and Mark Smith were passionate advocates of this view of Alzheimer's disease. The theoretical basis of the linkage was strong. Our brains are oxygen hungry. Even though they only weigh a little over three pounds (about 1.5 percent to 3 percent of a normal human's body weight), they use nearly 20 percent of the oxygen we breathe in. Using all that oxygen raises the danger that some of it will get loose and lead to cellular oxidation. We now also know that the cell uses oxidation and its reversal (called reduction) to signal among its various biochemical networks about changes in its chemical health. The oxidation hypothesis is that increased cellular damage caused by out-of-control oxidation is a big factor driving the losses in structure and function that cause the symptoms of Alzheimer's disease.

The correlation is strong, but the direction of causality is not. Does the oxidation lead to the damage we see in the Alzheimer's brain, or does some independent process that is part of the Alzheimer's disease program lower our defenses against oxidation? In other words, do we get Alzheimer's because we're rusting, or do we rust because we have Alzheimer's? As theoretically attractive as an oxidation hypothesis might be, because it is such a general phenomenon and a natural part of aging, it's hard to draw the threads into a clear pathway to Alzheimer's specifically. As its proponents pointed out, however, the specificity does not have to be there.

After all, if we were to come up with a drug that blocked the damage in Alzheimer's *and* Parkinson's disease, who's going to complain?

The supporters of the idea of a central role for oxidative damage in Alzheimer's disease were wary of the amyloid cascade hypothesis. It wasn't that they didn't believe that amyloid was present in the Alzheimer's disease brain. They just thought that it was working differently. The problem with the Aβ peptide, they hypothesized, was that it collected metal (copper, iron, etc.) when it formed aggregates and, in that form, drove an oxidation cascade, not an amyloid cascade. There were even those who thought that amyloid worked exactly backward from the way the amyloid cascade people thought about it. Because it could bind the metals, this group within a group thought that it would actually function as an *anti*oxidant. In this way, its presence was not a part of the Alzheimer's pathology; it was a sign that the brain was trying to defend itself against the stress of oxidation. As with the other groups, amyloid was not excluded from the thinking about the role of oxidation, it was just positioned in a different way in the disease pathway.

By now, I'm sure you're prepared for the warmhearted and open-minded reception that the Alzheimer's-is-oxidation hypothesis received in the writings of those who were promoting the amyloid cascade hypothesis. I'm kidding, of course. It received the same reception as did the inflammation hypothesis. Actually, a bit worse. They didn't even get a begrudging nod as a possible therapeutic approach to Alzheimer's disease. Again, using Hardy and Selkoe as an example, oxidation merits only a single mention in the last paragraph of the third section of the paper (possible therapeutic approaches). There were a total of six possible approaches listed. The sixth was described as a "broad amyloid-based strategy to prevent the synaptotoxic and neurodegenerative effects putatively triggered by Aβ." Trying antioxidants was one of three listed in passing along with the warning that "no slowing of cognitive decline has been documented in humans to date." Essentially, the partisans of the amyloid cascade hypothesis were strongly recommending against pursuing this or any other nonamyloid approach. What is amusing about their citing the failure of human trials to prove benefit is that today exactly the same acerbic comment could be applied to every one of the amyloid-based approaches.

OTHER OPTIONS: TAU AS A CAUSE
OF ALZHEIMER'S DISEASE

We return to the tangles in the Alzheimer's disease brain. Remember that Alois Alzheimer described two types of abnormal deposits in the brain of Auguste D.—amyloid plaques and neurofibrillary tangles. Because of the genetics of early-onset Alzheimer's disease, the plaques had become the main focus of the field. This was a powerful incentive to focus on amyloid, but there was never a disincentive to focus on tau. There were many in the field during this period who strongly supported the causative role of tau in the development of Alzheimer's disease. The arguments were based primarily on two observations. The first was that tau tangles were found in virtually all Alzheimer's disease brains. The second was that the brain regions where these tangles were found were much more logically tied to the functions that were lost in Alzheimer's disease. For example, there were plaques and tangles in the brain's memory center, which is affected badly during the course of Alzheimer's disease. Score one for both plaques and tangles. Some of the behavioral problems in Alzheimer's disease, however, are more closely linked to regions found lower down in an area of the brain known as the brain stem. Here there are no plaques, but the cells that are at risk have plenty of tangles. Tangles for the win.

This raises the question of who is driving the show in Alzheimer's disease. Do the plaques drive the tangles, or do the tangles drive the plaques? This question led to a heated debate in the field that took on the tone of a religious war, pitting those who thought that Aβ came first (the Baptists or βAptists) against those who thought the tangles came first (the tau-ists). Note that unlike the other alternative ideas that I've listed here, this one had the feel of an either/or situation. There was no way to stake out common ground, and defending one's amyloid or one's tau became a critical part of daily professional life. Looking back, even though in its time the question really did seem an important one to answer, a lot of heat but not a whole lot of light emerged from this debate.

The animosities that sprang up were really quite stunning during the period. In a way that I have never seen replicated in all my years in science, the frustration on both sides led to some pretty personal attacks. Even those attacks that were not personal were still more appropriate for a barroom than a scientific debate. I understand the frustration. As we have seen and will learn more about, the active suppression of all non-amyloid ideas was real. In my interviews with people who were in the field at the time, I've found that there were many who recognized what was happening but either couldn't or chose not to do anything about it. Nonetheless, it probably didn't help the cause to accuse people of praying at "The Church of the Holy Amyloid" or other such schoolyard taunts. It's funny in a way, but it was counterproductive. It simply validated to the amyloid cascade cabal[12] that all of the anti-amyloid arguments were being made by fringe elements. It delegitimized in many ways the honest attempts at holding the field to account and, in all probability, made matters worse.

SYNOPSIS

The elegant, almost magical, convergence of the genetics of Alzheimer's disease and Down syndrome with the biochemistry of the Aβ plaque peptide, the microscopy (neuropathology) of the Alzheimer's disease brain, and the biochemistry of the APP and presenilin proteins created a strong yet rational momentum toward the vigorous pursuit of the amyloid peptide as the one and only driver of Alzheimer's disease. The amyloid cascade hypothesis was the first consolidated exposition of these ideas and formed what remains to this day the predominant disease model of the Alzheimer's field. Unfortunately, that dominance was used to suppress many other worthy ideas, all of which were more or less compatible with the core ideas of an amyloid cascade. This took all of the diversity out of our research portfolio, blocked publication of studies into amyloid alternatives, and reduced funding for groups that wanted to explore other ideas. We will learn more about these negative consequences in the chapters to come. As you might imagine, the consequences of this dogmatic approach to research were devastating and, sadly, remain with us to this day. This is no way to study a human disease.

DOUBLE-EDGED SWORDS

We have seen that Alzheimer's disease is an incredibly complicated condition that doctors and scientists have only started to fully grasp. We have made some progress, of course, but our picture of the science of the disease is incomplete. To make sustained progress toward a real set of treatments, we have a lot of work to do. The effort is a large one, and it requires more resources than those we have talked about thus far. This next part of the book is devoted to looking outside the research labs of our universities and medical schools. Two other institutions are major players in the field, and both play important roles in setting the direction and the priorities of the search for successful Alzheimer's disease treatments. Both have had significant positive influences on the field, providing money and know-how to sponsor our research and help us move it from the laboratory bench to actual products that can be used in the clinic. As the title of this part suggests, however, both institutions have also held us back and in their own way contributed to the mess that we find ourselves in today. Each of the groups, therefore, is like a double-edged sword. One way it cuts for good; the other way it cuts for ill.

6

FEDERAL SUPPORT OF BASIC BIOMEDICAL RESEARCH

The work on Alzheimer's disease is not something that just happens. It takes hundreds of well-funded laboratories and thousands of well-trained professionals spending countless hours in their labs and clinics. An enterprise of this scale is not cheap. The question is "Who pays for it?" It probably won't come as a big surprise that in the United States and most other countries it's the federal government that pays the lion's share of the cost. Therefore, as a tax-paying citizen, that means that you are paying for it. You also pay for it through your charitable donations to nonprofit organizations such as the Alzheimer's Association, Cure Alzheimer's Now, Coins for Alzheimer's Research Trust, and many others. The scale of the financial commitment we citizens are making is impressive. It's worth taking a close look at the numbers to appreciate what great benefactors we have been.

Let's start by taking a moment to appreciate why basic scientific research is such an expensive undertaking. Science has advanced to the point where if someone like Leonardo da Vinci were alive today, he would not have the resources, the infrastructure, or the time to do the high-quality science he was able to do in the fifteenth and sixteenth centuries. In the twenty-first century, the costs of running a modern program of discovery in a single small research laboratory with only three people in it—a principal investigator, a student, and a technician—begins at around a quarter of a million dollars per year. Two-thirds of that cost, roughly, is needed just to meet payroll. The people doing the work are smart and well trained, and their toil does not come cheap.

The remaining one-third is for the resources to run the actual experiments. Let's use the DNA analysis done by the polymerase chain reaction, or PCR, as a practical example of why laboratory work is so expensive. PCR might be familiar to you because it appears as one of the methods used by the police in TV crime shows. The screenwriters tend to exaggerate a bit for dramatic purposes, but mostly they get things right. These days PCR is nearly ubiquitous in science. It's no surprise therefore that most Alzheimer's researchers make use of this technique. What the TV folks don't share with you, however, is how much it costs to equip your lab to run a PCR reaction. You need a thermocycler (~$3,000) and other small pieces of equipment (hundreds of dollars each). You also need "disposables" and one-time use items such as DNA primers and various enzymes. In the end, to equip your lab to run PCR, you will need to spend almost $16,000 in start-up costs plus running costs. In my own small lab, for example, three of us spend roughly $5,400 per year on just this one procedure.

In some ways $5,400 may not sound like a lot, but think about how many things you could buy with that money every year. Then consider that PCR is one of the least expensive procedures in a typical basic research lab. At the other end of the scale is the investment needed to purchase machines like DNA sequencers or high-end microscopes. Just for scale, an entry-level microscope with decent lenses costs around $35,000. Add the capacity to do fluorescence and take high-quality photographs, and the costs jump closer to $75,000. Higher resolution work on a machine known as a confocal microscope will run you north of $300,000. If you want to look at proteins with atomic resolution, you're going to need a cryoelectron microscope. That will set you back $10 million (not including running costs). You get the picture.

These are the kinds of dollars we need to meet our basic Alzheimer's disease research budget. And every year our representatives in Washington write a check. As taxpayers, we should stand proud of our long-standing support of basic research. It has required foresight, altruism, and a lot of patience on our part. That latter point is especially important because the payback we get from our investment is very indirect. Unlike clinical research that can turn a handsome profit for a drug company in just a few years, the "profit" we as taxpayers realize from our investment in basic research is less tangible and takes years or decades to realize. More frustrating still is that you really cannot draw a direct line from any given dollar's investment to its valuable return. And if you're doing financial analysis for a big pharmaceutical firm, this delayed-gratification approach is not going to make it out of the starting gate at decision-making time.

Of course, basic research can lead to profitable applied research, and there are many stories out there to prove that point. But it doesn't have to, and, to be honest, it usually doesn't. The path from discovery to application is winding and rarely logical. As a wise person once remarked to me, the light bulb was not invented by someone trying to make a better candle. And I would add that the LED was not invented by someone trying to make a better light bulb. A lot of seemingly irrelevant science was churned out over many years before people finally figured out how to make a practical light bulb. Great inventions seem nearly miraculous when we first look at them, but they do not come out of nowhere. In each case there is a foundation of basic science that triggers an insight in the mind of a clever inventor. Looking back, we can understand how some bizarre and seemingly trivial observation (a carbon filament glows in a vacuum when an electric current is applied) was put to very practical use (a light bulb) and made the inventor a lot of money. It all looks so simple in retrospect. But it cannot be stated too many times that it is the foundation of basic research—no matter how bizarre or seemingly trivial—that is the key to the progress. There are no shortcuts in building this foundation, and there are no crystal balls. We need to spend our research dollars in as many areas as we can. We should not just be investing in candle research.

Basic research is the key to advancing both science and medicine, and we, as taxpayers, are the ones who are paying for it. So, let's look at how our tax dollars funnel from us to the benches of the biomedical research laboratories of the world. In the United States, most of the research into human disease is sponsored by the National Institutes of Health (or NIH for short). The NIH is part of the US Department of Health and Human Services (HHS), a huge sprawling Department that includes the Centers for Disease Control and Prevention, the Food and Drug Administration (FDA), Medicare, and Medicaid, in addition to the NIH. As you can imagine, Medicare and Medicaid are expensive (over a trillion dollars proposed for the 2020 budget), but this type of funding is mandated by law. The annual "discretionary" HHS spending that needs to be reauthorized every year now runs between $80 billion and $90 billion. The NIH chews up a fair chunk of that amount.

Precise numbers can go up and down depending on the year. As figure 6.1 shows, there are periods of steadily increasing funding (1995–2005;

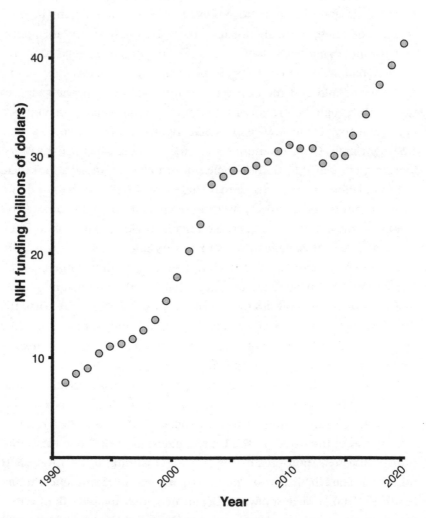

6.1   NIH funding during the years 1990–2020.
Source: Data are from https://www.nih.gov/about-nih/what-we-do/nih-almanac/appropria
tions-section-2.

2015–2020) and periods when growth is flat (2005–2010) or even decreases (2010–2015). But year after year Congress has seen fit to make a multibillion-dollar outlay for the NIH (the amount is scheduled to grow to $41 billion dollars in 2020). Even in today's world that is an extraordinary amount of money. To put that in perspective, that's almost as much as half the amount of money that the United States spends each year to fund veterans' benefits.

So, where does all that money go? The NIH budget is divided among 27 different institutes and six centers. There are institutes that cover entire fields (the National Institute of Nursing Research). There are institutes that cover specific diseases (the National Cancer Institute). There's even an institute for books (the National Library of Medicine). But how did this huge organization get started, and where does research into Alzheimer's disease fit in?

The origins of the NIH go back to the founding of the United States of America, but the multibillion-dollar enterprise we see today really only began to take shape in the middle of the twentieth century. The government's involvement in medical research dates back to 1798 when the Marine Hospital Service (MHS) was established to help merchant seamen with their medical needs. The role of the MHS expanded in the late nineteenth century as the health of arriving immigrants raised fears that they were importing diseases such as yellow fever and cholera. The MHS took a proactive approach to these concerns. In 1887, one of their staff physicians, Joseph J. Kinyoun, established a small laboratory on Staten Island in one of the MHS hospitals. Kinyoun called this facility a "Laboratory of Hygiene" and based it on new ideas about infectious diseases from Europe. He had soon identified the bacterium that causes cholera in several MHS cases, and thus, nearly 100 years after its founding, the MHS demonstrated that government investment in basic research could lead to improved public health not only for sailors but for anyone threated by infectious disease.

Fifteen years later, Congress agreed to construct a building to house Kinyoun's laboratory in Washington, DC. This was the humble beginnings of the NIH. Even so, Congress was not fully on board with these expenditures in public health research. As the NIH website wryly notes,

The founding legislation for the NIH . . . resides in a routine supplemental appropriations act. Many other scientific agencies of the federal government were also created via [similar] "money bills." Congress was not convinced that such bureaucracies would prove demonstratively useful, so it chose to preserve the option of divesting the government of them simply by not renewing their funding. (https://history.nih.gov/exhibits/history/index.html)

Looking back, as the graph of the budget of NIH expenditures shows, the ensuing 100 years has removed any doubt from the minds of members of Congress that this investment is "demonstratively useful."

The value of basic research in this larger public health effort was made explicit in 1902. That year Congress reorganized the MHS and renamed it the Public Health and Marine Hospital Service (PH-MHS), a precursor to what is known today as the Public Health Service. As part of this restructuring, a research program was established, and for the first time Congress allowed the PH-MHS to hire individuals with PhD degrees to head the funded research teams. This decision seems logical in hindsight. But allowing someone besides a medical doctor to run a laboratory devoted to human health was a critical broadening of the original conception of the MHS. This and other actions established that the government valued health-related research sufficiently to put real money behind the enterprise. But it was only after the Second World War that these early embryonic forms of the NIH grew into the agency we recognize today. The National Cancer Institute came first. It piloted a program that allowed it to fund research outside of its Washington laboratory. The experiment proved successful, and for congressional representatives, bringing federal research dollars to local institutions proved hugely popular. As a result, the NIH budget grew from $8 million in 1947 to more than $1 billion in 1966.

The story of each institute of the NIH is its own lesson in the politics of science and medicine. But we want to know about research into Alzheimer's disease. One of the 27 institutes that would seem well suited toward engaging in Alzheimer's research is the National Institute of Neurological Disorders and Stroke (NINDS). It was established in 1950 (originally the National Institute of Neurological Diseases and Blindness) with the goal of understanding the neurological and psychiatric problems faced by soldiers returning from World War II. NINDS is where problems of brain health are studied, and there are robust programs in areas such as developmental and genetic disorders, epilepsy, movement disorders such as amyotrophic lateral sclerosis (ALS, also known as Lou Gehrig's disease) or multiple sclerosis, and late-life degenerative disorders such as Parkinson's disease. This seems an obvious organization to run an all-out assault on Alzheimer's disease, and yet this is not where the bulk of the NIH funding for Alzheimer's research comes from.

A second agency that could serve as a possible source of funding for the work is the National Institute on Aging (NIA). The logic for running Alzheimer's disease funding out of the NIA is a bit indirect. Nonetheless, as we will see in chapter 10, a good argument can be made for grounding Alzheimer's basic research in the biology of aging. By this logic, while not perhaps the most obvious choice, we may argue that the decision to use the NIA as the vehicle to fund Alzheimer's research is at least defensible. Of course, a number of other neurological diseases are also age-related— Parkinson's disease, ALS, Huntington's. Yet these diseases are primarily funded through the NINDS, the agency explicitly devoted to the study of the brain and its diseases. Why, you may ask, is Alzheimer's disease singled out for funding by the NIA? The answer is really quite simple: politics.

The NIA is relatively young. Units devoted to the study of aging had been a part of other federal organizations, including the National Institute of Mental Health (NIMH), but it was not until 1974 that a separate organization, the National Institute on Aging, became the eleventh institute of the NIH. Vying for the attention and dollars of Congress, the young institute struggled for recognition. It was years after its establishment that Robert Butler, the leader of the NIA, hit upon a strategy to increase its visibility with the public and, more importantly, with the members of Congress he was hoping would fund NIA. He was joined in this effort by other early thought leaders including Robert Terry, Robert Katzman, and Zaven Khachaturian. These men realized that while aging is important and interesting, it isn't something we can avoid or "cure." We can all agree that cancer is bad and that we should be willing to invest money to get rid of it. The connection is so intuitive that one need look no further than the proclamation by President Richard Nixon of a "War on Cancer" to see the strategy at work. A war on cancer might seem a bit hyperbolic. After all, how do you go to war with a disease? On the other hand, it does have a nice public relations ring to it.

Aging is different and much more of a public relations challenge. None of us wants to age, but it doesn't seem likely that we can stop it, and it would certainly seem foolish to declare war on it since victory might be elusive if not impossible. Aging, according to the logic of the early leaders of the NIA, doesn't have the same marketing glamour to it as cancer. But, they asked, what if we could find a disease that was an unavoidable part of aging, a disease that people were really afraid of? That fear would engage the public, capture Congress's attention, and bring federal research money to the field. Alzheimer's disease fit the bill perfectly. The public relations challenge could be met by changing how "aging" was framed.

As we saw in chapter 2, Kraepelin shaped the narrative surrounding Auguste D.'s rare dementing illness by calling it a specific disease, Alzheimer's disease. In doing this, he realized his goal of shaping the ways in which future physicians and scientists would view and talk about late-life dementias. In a similar way, the Butler-Khachaturian-Katzman-Terry marketing strategy to develop support for the young NIA required a new reshaping of the narrative. Alzheimer's disease needed to become recognized as a most common form of dementia.

For their strategy to work, the public and their representatives in Congress needed to be convinced that Alzheimer's was prevalent, malevolent, and costly. That was not how most physicians viewed it in the 1960s. Alzheimer's disease was still viewed as a rare, atypical type of old age dementia. As with Auguste D., it was rare to see patients in the clinic with a full-blown dementia before they were in their seventh or eighth decade. Sure, as we got older there was cognitive decline. At the time, the process of aging was seen as being accompanied by a natural slowing of mental capacity—a hardening of the arteries as it was sometimes called. This was normal and common; Alzheimer's disease was seen as relatively rare.

For the NIA leadership, this viewpoint had to change. Applying the principles of marketing, they developed a two-pronged strategy. The first goal was to make Alzheimer's disease much more sinister, a fairly easy sell. The second prong of the new approach was an energetic push to greatly expand the meaning of "Alzheimer's disease." The idea was to use it as a label for as much old age dementia as possible. The NIA reached back to the ideas of Kraepelin and Alzheimer that the deposits of amyloid plaques and neurofibrillary tangles were the true cause of the disease. For the young NIA this amyloid-is-Alzheimer's approach turned out to be a gold mine. As more and more cases of dementia were analyzed under the microscope by neuropathologists, more and more examples were found where the autopsy of a person who had died with a serious clinical dementia revealed amyloid plaques and neurofibrillary tangles in a characteristic distribution in the brain. Well, they said, this must be Alzheimer's disease. The once rare form of presenile dementia had now graduated to a disease that explained a huge fraction of all age-related cognitive decline—60 percent by some estimates. Looking back on this history, it is noteworthy that the success of the NIA leadership in redefining dementia represented the second major inflation of the boundaries of what we call Alzheimer's disease. Even more than the early efforts of Kraepelin, this inflation was strategic and political, not scientific.

The strategy worked really well. Money and other resources were going to be needed to deal with this terrible disease of previously unrecognized prevalence. Sure enough, the federal investment in Alzheimer's disease research began to increase. One notable milestone occurred in 1984 when the first Alzheimer's Disease Research Centers were established under the auspices of the NIA. These were large grants awarded to famous researchers at prestigious US medical schools. They were trumpeted by the NIA as evidence of its commitment to defeating this horrible (albeit newly defined) monster. As the years went by, Alzheimer's research became a larger and larger part of the NIA research budget. So successful was the strategy of the second inflation that, as of today, the NIA spends about two-thirds of its entire $2.6 billion budget on Alzheimer's disease research. That's a lot of money to be spending on one disease, and it naturally raises the question of whether the Butler-Khachaturian-Katzman-Terry strategy was too successful.

One way to look at this is to say that even though $1.5 billion is a lot of money, we might well want to invest even more. This argument is based on the calculation that the returns on our investment will vastly exceed its initial costs. The math is pretty straightforward. The entire NIA has a budget of $2.6 billion. Recall that Alzheimer's disease costs the US health care system (in other words, you and me) nearly $300 billion each year. If we could reduce that amount just by half, the savings in a single year could fund the entire NIA for over 50 years. In other words, we could invest in the NIA at current levels for over 50 years, and that investment would pay for itself in a single year even if we were only partially successful in reaching our goal to cure Alzheimer's. Pass me my checkbook, please.

The problem is that there is a second edge to this multibillion-dollar sword. And if you were wondering why I bothered with a long description of the history of the NIA, it is here that you can find your answer. The problem is that the strategy for growing the federal investment in Alzheimer's has been so successful that now the tail is wagging the dog. Recall that the NIA was founded on a commitment to improve our understanding of the biology of aging and all of its related diseases. The premise was and still is that age contributes meaningfully to diseases that affect virtually every organ in our body. The original idea was that if we could understand the aging process, it would impact an entire range of human diseases. But aging alone, as a "product," lacked the curb appeal of cancer or allergy or infectious disease. So, to make aging an enemy worthy of attack and a project worthy of the investment of federal dollars, Alzheimer's disease was conflated with aging and inflated to help sell the idea of studying aging. But when Alzheimer's became the face of aging, it also became the face of the NIA. And increasingly, attracting money to aging means attracting money to Alzheimer's. This is backward from the way we ought to be approaching the problem, and this is the second edge of the sword. From this perspective, we should not so much be cutting back as rebalancing and repurposing how we spend our dollars.

One way to see the extent to which we are currently out of balance is to note that of the 43 milestones that the NIA lists as significant since the year 2000, over one-third are related to Alzheimer's, dementia, or cognition. Is one-third of the problem of aging really related solely to mental capacity in general and Alzheimer's in particular? I would argue that this number represents a serious distortion of the core mission of the NIA. This is particularly true because there are two other institutes at NIH— NINDS and NIMH—where work in this area can (and does) go on. This should free the NIA to focus more resources on the biology of aging, but it does not seem to have worked that way. The tail is wagging the dog.

Another way to look at the power of the Alzheimer's tail over the NIA dog is to look at the way in which the NIA budget is spent. The table in figure 6.2, from the NIA website, shows the breakdown of the budget by broad category of research. The 2018 numbers are about money actually spent, so they are the most accurate. The category of "Neuroscience" accounts for about 65 percent of the total NIA budget (see figure 6.2).

	FY 2018 Actual FTEs	FY 2018 Actual Amount	FY 2019 Enacted FTEs	FY 2019 Enacted Amount
Extramural Research:				
Aging Biology		$262,439		$317,816
Behavioral and Social Research	-	305,613		370,100
Neuroscience		1,575,481		1,907,922
Geriatrics and Clinical Gerontology		215,156		260,556
Subtonal, Extramural		$2,358,689		$2,856,394
Intramural Research	260	$148,566	250	$156,209
Research Management and Support	156	$64,248	185	$70,808
TOTAL	416	$2,571,502	435	$3,083,410

6.2 National Institutes of Health budget authority 2018 and later estimates (in thousands of dollars). Note that neuroscience is by far the largest number in the table. *Source*: Data are from https://www.nia.nih.gov/about/budget/fiscal-year-2020-budget /fy-2020-amounts-available-obligation.

Most of that money went to Alzheimer's disease or dementia research. And even that is a bit of an underestimate of the slice of the pie devoted to Alzheimer's. The "Behavioral" part of the category of "Behavioral and Social Research" includes a lot of dementia research. Plus, if you look at the laboratories that are at the NIH itself (the Intramural Research Program), you'll find that some of that research is also related to dementia or Alzheimer's disease. All told, it would appear that over two-thirds of the NIA budget goes to Alzheimer's research, either directly or indirectly.

It needs to be said that there is a defensible rationale for favoring dementia research with more than its proportional share of the dollars. All of our body's systems deteriorate with the passage of biological time. As we age, we are at increased risk for an entire menu of problems we would love to avoid. Our muscles get weaker, our bones get more brittle, our heart function declines, our immune systems become less responsive, our risk of cancer goes up, and dementia becomes more and more likely to strike. Solve the puzzle that is aging, and we would have a huge positive impact on all of these conditions. But aging research is pretty much in its infancy. It will likely be many years before basic research offers us practical tools that we can apply in the clinic. The public and their representatives in Congress are constantly asking, what can we do *right now*? How shall we prioritize the spending of our resources? Where shall we start our attack on this complicated problem?

I was once privileged to be a part of a workshop whose purpose was to bring together geriatric physicians and basic scientists. As we were wrapping up, I asked the physicians in the room the following question:

We've spoken about many issues today—diabetes, sarcopenia, dementia, falls, infections, and others. My question to you is, as practicing physicians, if I were to give you a magic wand that you could wave and remove one, but only one, of the items on the long list of problems we've been talking about today, which one would you choose?

There was not even a moment's hesitation before every single one of them said, "dementia." When I asked them why, they were very clear. For all of the other conditions there was something that they as physicians could offer the patients and their families. They could treat diabetes with diet or drugs. They could replace knees or hips. They could prescribe statins to lower cholesterol. They could offer exercise programs to improve muscle tone or balance. But there was nothing that they could do for dementia. And not only that; if a person came to them with dementia, treating everything else—their diabetes, their motor coordination, their arthritis—became much harder from a clinical point of view and more vexing from an ethical point of view.

Given the urgency of the problem of dementia as expressed by these physicians, one begins to appreciate the attraction of focusing heavily on Alzheimer's disease and of putting most of the NIH funding eggs into the Alzheimer's basket. It may be lopsided, but it is at least something that we can do right now. This is exactly the situation that the NIA leadership faced in the 1980s. The long-term benefit of building a solid foundation of basic aging science that will one day spawn a new generation of therapeutics is hard to sell. We, the public, want answers sooner rather than later, and, of all the problems wherein basic research might make a difference, dementia is a compelling place to start. In this context, then, a certain amount of nonuniformity in funding allocations makes sense.

However, an equally strong case can be made for the idea that if you've put most of your eggs in the Alzheimer's basket and you trip while you're walking home, you've got nothing to use to make dinner. That is where we find ourselves today. The Alzheimer's field has tripped, and we have hardly any eggs left in our aging basket. Some of the distortion of funding allocation made sense, but using two-thirds of the entire budget—year in and year out—does not.

This is not just an NIA problem. The NIH spends a lot of money on dementia, and as I've suggested, there are other institutes at the NIH that fund Alzheimer's research. Their contributions are small compared to the NIA, but the numbers are still pretty impressive. The NIH identifies about 3,300 extramural grants that fund research into Alzheimer's disease and related dementias. To appreciate the extent to which NIA dominates Alzheimer's disease research, consider that 90 percent of this total (about 2,950) are grants from the NIA. The bottom line is that Alzheimer's research drives the NIA, and the NIA drives Alzheimer's research. This is a lopsided approach to dealing with Alzheimer's disease, and it is certainly off-kilter with respect to the study of aging as a contributor to human health problems. It takes resources away from studying the root of the problem, namely, the biology of aging itself.

The distortion of mission gets worse when we consider the specific Alzheimer's research topics that are funded by the NIA. We learned about the amyloid hypothesis earlier and the factors that drove it to dominate the thinking in the field, culminating in the advice, "If you aren't studying amyloid, you aren't studying Alzheimer's." The result of this laser-like focus is not only does Alzheimer's research drive the NIA, the amyloid cascade hypothesis drives Alzheimer's disease research. One way that one might hope to illustrate this point would be to simply ask how many NIH grants are based on the amyloid cascade hypothesis. But identifying the number of amyloid dollars is more than a little tricky; even people at the NIA have a hard time coming up with a realistic exact number. The hypothesis that amyloid and plaques are the causes of Alzheimer's disease is so deeply ingrained in the Alzheimer's field that it is very hard to tease apart what fraction of the grants awarded by the NIH are based on that theory.

The NIH has a software tool, called the RePORTER (https://projectreporter .nih.gov/reporter.cfm), designed to search for words in the text of funded NIH grants. If we search on "Alzheimer's disease," we max out the system at 500 grants (RePORTER stops looking after it finds 500 "matches"). This is, of course, only about 15 percent of the 3,300 total Alzheimer's grants, but even this simple exercise underscores my earlier point about how Alzheimer's drives the NIA. The funding for 497 of these top 500 Alzheimer's grants came from the NIA. By contrast, if we search on the text "Parkinson's disease," only 51 of the top 500 grants are NIA funded; 309 are funded by NINDS. Even if we search "not amyloid" and include words like "metabolic and mitochondrial abnormalities" in addition to "Alzheimer's disease," the top 500 grants are still more than 80 percent NIA funded. And even though I had specifically tried to find grants that did not focus on amyloid, 7 of the top 20 grants from the NIA still had the word amyloid in the title in a positive way.

Another measure is the NIA sponsorship of the much cited "NIA-AA Research Framework."[1] This document is and will continue to be a touchstone for Alzheimer's disease research for years to come. As we will see in more detail in later chapters, to say that its recommendations are amyloid-based would be a serious understatement. The following statement is a direct quote from the paper's abstract:

Although it is possible that β-amyloid plaques and neurofibrillary tau deposits are not causal in AD pathogenesis, it is *these abnormal protein deposits* that *define AD as a unique neurodegenerative disease* among different disorders that can lead to dementia [emphasis is in the original].

The translation of this sentence is a repeat of the advice our Alzheimer's group received in the 1990s: "If you're not studying amyloid, you're not studying Alzheimer's." The fact that the paper was written with the sponsorship of the NIA makes it clear how much the NIH sees the biology of Alzheimer's disease through an amyloid lens. As I have tried to stress, amyloid is not irrelevant, but it cannot and should not be used to define Alzheimer's disease.

The influence of the NIA on funding decisions extends well beyond the NIA itself. Consider that the "AA" in the title of the research framework stands for Alzheimer's Association. This nonprofit organization began in 1980 as a research and patient advocacy group and has since grown into the largest nongovernmental funding source for Alzheimer's disease research. Internationally, they fund hundreds of millions of dollars' worth of research. Of course, if your name is "Alzheimer's Association," one expects that your research dollars will go to funding Alzheimer's disease research almost exclusively. One does not expect, however, that the bulk of that funding will go to a single amyloid-based approach as it has historically at the association. While this emphasis has begun to change, other smaller, organizations such as the Alzheimer's Drug Discovery Foundation, Coins for Alzheimer's Research Trust, Clear Thought Foundation, and others have led the way in promoting a far more diversified approach to Alzheimer's disease.

The dominance of the amyloid hypothesis as a disease model has had many chilling effects on the research efforts to find effective treatments for Alzheimer's disease, but as these examples indicate, nowhere is the chill more Arctic in its effects than in the distortion it has placed on the funding of dementia research. The government, through the NIH, picks up most of the basic research bill. Thus, with amyloid clinical research so well covered by industry-sponsored research, one might have hoped that the NIA would take the opportunity to lay a large and broad foundation for subsequent generations of post-amyloid Alzheimer's research. Yet this is not what's happening. The NIA funds a lot of good research into aging. But the 800-pound gorilla in the funding portfolio is Alzheimer's disease and the role of amyloid and tau. The gorilla has cast its long shadow over the funding decisions related to other strategic approaches. To stretch the analogy a bit, the amyloid flea is wagging the Alzheimer's tail which is wagging the NIA dog.

This is not how you study a human disease.

7

THE PHARMACEUTICAL
AND BIOTECH INDUSTRY

While the pharmaceutical industry relies heavily on a vibrant community of scientists engaged in basic research, their financial support of the overall basic research effort is shrinking. Instead, the companies collectively represented by the Pharmaceuticals Research and Manufacturers of America, known as PhRMA, are more and more heavily focused on clinical trials. If you count the costs of clinical trials as research, there are studies out there that estimate industry outspends the government by nearly two-to-one.[1] That's research for sure, and the effort sounds laudable. But these dollars are going overwhelmingly to what we call "applied" research. Industry is willing to spend lots of money on trials because the financial payoff is relatively quick (years, but not decades). When a pharmaceutical company starts a new clinical trial, they are betting that they are going to make a fortune when the drug that they are testing turns out to be the next blockbuster. From the standpoint of the accountants in the front office, the decision to run the trial is risky, but the risk-benefit ratio is favorable. These same accountants, however, are going to be much less charitable about putting company dollars into basic research. The timeline for a return on the investment is long and thus makes little short-term financial sense for the company.

With PhRMA focused more and more heavily on clinical trials, the responsibility for funding the basic science has fallen on the shoulders of investigators at universities and research institutes around the world. This is not to ignore that there are dedicated scientists at small-to-medium biotech firms that contribute as well, but they need to protect their own intellectual property. To make their backers happy, their priority has to be return on investment, not scientific knowledge. That puts them in a position where they cannot consistently publish cutting-edge basic science that advances the field.

The relationship of the pharmaceutical industry to Alzheimer's disease is a complicated one. Of all of the double-edge swords, this is the one whose two blades are the sharpest. That two-sided nature of the industry contribution to our current situation made this chapter among the hardest for me to write. I wanted no villains in this book, but my first instinct was that the pharmaceutical industry came close. There is certainly plenty of evil in the industry. To find it, one need look no further than the unbridled greed that contributed to the opioid crisis or to the profits made on common drugs such as insulin. The hypercompetitive, cutthroat nature of the industry, with its focus on market share rather than human health, seemed well suited to serve as a whipping boy for our discussion of the state of Alzheimer's disease research. No matter how compelling PhRMA might look as a target, however, the actual role of the industry in the Alzheimer's disease story is unexpectedly different.

In the quest for an Alzheimer's drug, there is no Snidely Whiplash character whose story serves as an exemplar of how not to study a human disease. The culprit in the end is an amorphous decision-making process in the laboratories and boardrooms of the pharmaceutical industry that is not so much evil as gullible. I expected to find evil because of stories such as the ones surrounding OxyContin and insulin. I found that the story of Alzheimer's drug development is different. Both OxyContin and insulin were established products. There was no need for research, product development, or regulatory approval. The industry was simply trying to sell us larger and larger quantities of two well-known and well-liked drugs and was doing its rapacious best to squeeze us consumers for every last dollar. We may wince, and we can discuss the morality of these decisions, but basically that's just business doing business. The problem with Alzheimer's disease for the industry has been, and continues to be, that there is no product to sell. They may salivate over the *potential* of the Alzheimer's disease market, but right now the marketing folks have nothing to work with. They just sit and wait.

In the end, it turns out that industry's role in the situation that we find ourselves in today is best seen as a story of incompetence. Their main contribution to the current state of Alzheimer's disease research is not what they did, but what they didn't do. They didn't listen to their own data and ended up making not just bad scientific decisions, but bad business decisions. To put this story in context, it helps to understand where the firms we now recognize as pharmaceutical giants got their start and how they grew to their current size and power.

HISTORY OF THE PHARMACEUTICAL INDUSTRY

The drug companies of today all started from rather humble beginnings. Each story is different, and yet there are many common features. Bayer AG is one example. The company was started in 1863 by Friedrich Bayer, a salesman, and Johann Friedrich Weskott, a chemist who specialized in fabric dyes made from coal-tar derivatives. They invested in research with the goal of creating new, better, and cheaper dyes. They also started using some of the same production techniques to turn out pharmaceutical products. Salicylic acid, modified by the addition of an extra acetyl group, produced acetylsalicylic acid, a product we know as aspirin. Bayer chemists didn't invent this process, but they began to manufacture and market it once they proved it had pain-relieving properties. They won a trademark for the name "Aspirin" and marketed it very aggressively. Bayer turned aspirin into what is still viewed as one of the wonder drugs of history. That mix of chemistry and salesmanship was the basis of the original partnership. The fact that the company was named after the salesman (Bayer) and not the chemist (Weskott) we can take as a bit of foreshadowing. Today Bayer is among the biggest of the drug companies with annual revenues of around $40 billion. They no longer just make pharmaceuticals. They've expanded to consumer products and also have a presence in animal health and crop science.

Other pharmaceutical giants began their lives in areas more closely related to where they are today. Merck began its rise to the fifth largest pharmaceutical company (based on 2018 revenue) with early work in vitamins.

Eli Lilly (now the seventh largest) began its rise by manufacturing and selling quinine; Takeda, an important player in Alzheimer's disease drug development, began in eighteenth-century Japan as a distributor for traditional Eastern medicines, and in the nineteenth century began distributing Western medicines as well. Johnson & Johnson, by far the largest of today's pharmaceutical companies, with annual revenues approaching $400 billion, began as three brothers peddling antiseptics and surgical dressings.

There are similarities in these origin stories. A few companies started in chemistry, others began in manufacturing, and others in distribution, but in very few cases did the founder or group of founders take an original observation and translate it into a profitable health-related product. In reality, the key to the success of each company was first and foremost the marketing skills of original players. Business acumen mixed with no small amount of raw ambition combined to make these companies what they are today. This was true in their beginning and also at many watershed moments along the way. Most of them recognized the value of scientific research as a source of new products, but research was not what made the company tick. The Bayer AG website says it well. In describing the early days of Bayer and its competitors, they note that "only innovative companies with their own research facilities *and the ability to exploit opportunities on the international market* managed to survive over the long term [my emphasis]."

This mix of science and salesmanship is the Central Dogma of the Pharmaceutical Industry. It is PhRMA's DNA double helix. And it is our double-edged sword.

WHY DO WE NEED THEM?

The sales and marketing side of PhRMA is ever present in our world. It ranges from the mere annoyance of TV ads for drugs that deal with this or that medical condition to ethical nightmares such as the Sackler family's role in the opioid crisis and the pricing practices of Martin Shkreli. This take-no-prisoners approach to pushing drugs is costly for us as health care consumers, and it begs the question of whether we really need big companies making our drugs for us. As the National Academies of Sciences, Engineering, and Medicine concludes in a recent study,[2] "Drugs that are not affordable are of little value. . . ."

The problem is this: we really do need the pharmaceutical industry. The quote above is only half of the Academies' conclusion. The full quote states that "drugs that are not affordable are of little value and drugs that do not exist are of no value."

The last part of that quotation is the other edge of the PhRMA sword. Tomorrow, if I were to walk into my laboratory and discover the cure for Alzheimer's, I've not done much to help ease the public health burden of dementia in our society. If there were no one to turn my discovery into a product and do it at scale, I would be talking to myself and maybe a few of my academic friends. At a practical level, my discovery would be worthless. Some company would have to take my idea and "reduce it to practice." Without that, a drug would not exist "and drugs that do not exist are of no value."

Bringing a concept to market is a long and arduous journey, and it is not for the faint of heart. Suppose my fantasy discovery was that an extract of leaves from the weeping willow tree in my backyard removed plaques and tangles and halted brain deterioration in two different mouse models of Alzheimer's disease. Fantastic, but curing mice is not the goal. We need to take care of 50 million people worldwide who currently have Alzheimer's disease and many millions more who are likely to get it. My backyard weeping willow extract is only a first step. Having found an activity, the next step would be to use analytical chemistry to find the active ingredient. Then we would need synthetic chemists to tinker with the structure of the active compound to find out whether modifications might make it more potent. Aspirin, you'll recall is a modified form of sal-icylic acid, the original pain-relief compound. While all this is going on, we would need to bring in the patent lawyers so that every compound—the parent and all its derivatives—would be covered by patents. Ideas are notoriously slippery; a patent provides a solid and legal product that can be more easily bought and sold. Next, we would need to do what prac-titioners call pharmacokinetics. These types of studies answer important questions such as how long the compound stays in the body, whether it gets into the brain, how often it has to be taken, and how much needs to be taken at a time. Once we cleared these hurdles, we would need to start human clinical trials. Phase I trials test for safety, Phase II trials test for effectiveness in small numbers of people (dozens), and Phase III multisite studies test for effectiveness in larger numbers of people (usually a thousand or more). These trials are incredibly expensive. Phase III Alzheimer's disease trials, for example, regularly cost hundreds of millions of dollars each. If all goes well, the next step is to ask for regulatory approval to sell the product. This step will differ from country to country but is critical for commercial success. With approval from agencies such as the FDA, engineers can go to work designing the manufacturing procedures that would be needed to produce the product at a truly industrial scale. The distribution team would need to work out the logistics of getting the manufactured product to all the pharmacies of the world. Then, finally, you can call in the marketing team and begin to promote the product.

The list of steps needed to go from the discovery of a potential Alzheimer's disease cure in my backyard to a product that the people of the world can use is long, and the costs associated with getting through the list run into the billions. Not only is the process complex and costly but it is agonizingly slow. Estimates of the length of the drug development timeline range from 10 to 15 years from discovery to product . . . and that's if everything goes well. One final, painful detail needs to be added. No matter the rigor and care with which the original research is done, fewer than one in ten drugs that start Phase I clinical trials end up with FDA approval. That means that the green light to market a new drug and start recouping the enormous up-front investment is never achieved for nine out of ten products that start the process. This is not a project for the risk averse.

The drug development process is so long and so expensive and so risky that only a very substantial organization can muster the resources to see it through to its end. In other words, if we want a new drug for Alzheimer's, we need PhRMA. No academic can do it on his or her own. No university or research institute is secure enough to take on the risk. Even a small-to-medium start-up company would have a tough time running the gauntlet of barriers put up in the way of bringing a product to market. Giant corporations such as the companies listed above are needed to translate a good idea into a real product. Bottom line: for all of its excesses, if PhRMA did not exist, we would have to create it.

OK, but here we are twiddling our thumbs waiting patiently for a new Alzheimer's disease drug. Couldn't we speed things up a bit? That's a great question, and many smart people have looked at the time and resources involved and asked how we might hope to streamline the process. One approach would be to speed up the clinical trial process. A second would be to reexamine the regulatory criteria that must be met before a drug can be marketed. A third would be to alter the way in which the industry approaches the problem. Certainly, with a new case of Alzheimer's disease every minute in the United States alone, there is a lot of pressure to move things along more quickly. So, what is holding us back? We will deal with the clinical trials and regulatory process in the next chapter. For now, our focus is on how PhRMA does things.

THE INDUSTRIAL APPROACH TO RESEARCH AND DEVELOPMENT

The pharmaceutical giants all recognize the value of basic research. The problem is that they recognize it in the same way that Winston Churchill was said to recognize the value of vermouth in his martini: fill the glass with gin, bow to the vermouth in the other room, then enjoy the cocktail. In drug development the analogy is pretty direct. Most of the majors fill their glass with marketing strategy, bow to the research effort in the other room, and enjoy the profits. Yet while one can enjoy a glass of gin without vermouth, there is no way to enjoy the profits from a new drug for Alzheimer's disease without basic research.

That shouting you hear in the background is the sound of industry loudly calling foul. They point to data such as those compiled by the National Science Foundation (NSF), that show it is industry pays its fair share of basic research (see figure 7.1). According to these figures, in 2015 the R&D dollars contributed by industry constituted an impressive share of the total amount.[3] This was especially notable as it marked a crossing of the contribution curves for the first time in history. It bears noting that NIH funding has increased significantly since these data were compiled, but all well and good. This level of analysis however misses the question of how much of industry's contributions represent truly "basic" research. Clinical trials have become much more expensive, and they are important parts of the drug development process, but their purpose is not to gain new knowledge about disease mechanisms. Their purpose is to test safety and efficacy of already known quantities.

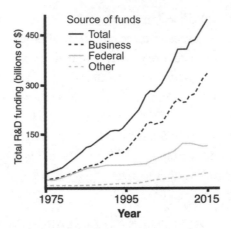

7.1 Distribution of research funding support.
Source: Data are from National Science Board | Science & Engineering Indicators 2018.

Determining how much drug development costs contribute to the total is surprisingly hard. The companies tend not to break down their research expenditures in a very granular way, and that makes it difficult to know how the funds listed as "research and development" are spent:

... "quite a lot" of things can be incorporated into R&D expenditures, such as legal expenses for acquiring and defending intellectual property rights, commercial activities, and fees paid to doctors for participation in clinical trials.[4]

That means that the idea that industry contributes more than half of the cost of true basic research is unlikely. A GAO report to Congress from 2017 (GAO-18-40) estimates that "from 2008 through 2014, worldwide company-reported R&D spending, . . . [mostly] . . . went to drug development (rather than research)."

This same report makes an effort to separate basic research from development. The results paint a very different picture that highlights the outsized role of federal funding in sponsoring basic research (see figure 7.2). These are 2014 numbers, but there is not much year-to-year change. From these data it's clear that industry spends a lot of money on research, but most of it is directed toward applied research and development, not basic discovery.

A separate study analyzed the 2017 portfolios of Pfizer and Johnson & Johnson.[5] The authors took the most profitable drugs sold by the two

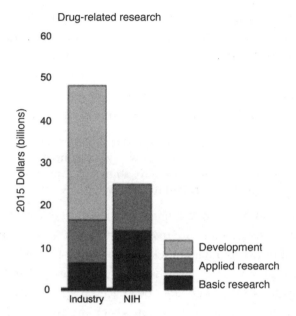

7.2 Comparison of federal- and industry-sponsored drug-related research.
Source: GAO analysis of NSF Business Research, Development, and Innovation Survey Data I GAO-18-40.

companies and worked backward to figure out where the foundational research (the equivalent of my backyard weeping willow research) had been done. Of the 62 products they analyzed, only 12 (less than one-fifth) were discovered by the companies themselves. The remaining 81 percent of the drugs all had their conceptual origins elsewhere. Four discoveries were acquired from others for every one discovery that was made inside the company.

There was a time when industry sought to change that 4:1 balance. In 1967, for example the pharmaceutical company Roche (Hoffmann-La Roche) established the Roche Institute of Molecular Biology in Nutley, New Jersey. The institute was devoted entirely to research and turned out a lot of well-regarded work. Unfortunately, the hopes of the company that they could channel the institute's basic research in a direction that would profit the company faded. The institute lasted for a mere 28 years before being summarily closed, forcing many well-known scientists to scramble, looking for new jobs. That mercurial, unpredictable feature of industry's commitment to basic research is evident even when it is being done in-house. For reasons best known only to themselves, upper management will suddenly upend an entire line of work. One particularly notable example is the story of Pfizer's commitment to Alzheimer's disease research. In early 2018 Pfizer management took a look at the drug development landscape and made the corporate decision to immediately stop work not just on Alzheimer's disease but virtually all neuroscience research:

As a result of a recent comprehensive review, we have made the decision to end our neuroscience discovery and early development efforts and re-allocate [spending] to those areas where we have strong scientific leadership and that will allow us to provide the greatest impact for patients.

The cynical among us might question the extent to which the decision was guided by concern for patients, but whatever the corporate logic, the result was hundreds of talented scientists abruptly laid off.

My point is not to argue that the resources industry spends to develop a new drug after its discovery are not a necessary part of the global effort to fight disease. My goal is to take a step back and consider the process in its entirety. In the path from bench to bedside that I charted a few paragraphs ago, the basic research component stops at or near the search for chemical modifications of the parent compound that might be better or safer. One could quibble about whether pharmacokinetic studies should also be considered basic research, but it's plain that all of the other steps in the process represent product development; the process of discovery is complete. The investment needed to turn my weeping willow leaf extract into a product would be huge, and the rollout of the product would take years to complete. I could not do it on my own. Nonetheless, in this fictional example, there is a germ of a larger truth: the discovery on which the entire effort is based remains my original basic research findings. Without that work, all the money and industrial might of PhRMA would not have been worth the spreadsheet it was tallied on: ". . . a drug that does not exist is of no value."

OUTSOURCING

The industry is not blind to the importance of basic research. Read their mission statements or watch how they promote themselves in TV ads and on the internet and you can see their view. But the lesson of the Roche

Institute has never been lost on the managers of these huge firms. They see it as an object lesson of how 28 years of investment in cutting-edge research doesn't do much, if anything, to plump up the bottom line. Despite its unquestioned scientific successes, in a financial sense the Roche Institute of Molecular Biology was an expensive failure. This has led to much clucking of accountant tongues and mutterings about pulling up stakes and getting out of the basic research business. Let someone else do the early work. This makes good business sense, and so the industry has evolved away from basic research.

There's a problem, though. This business-oriented approach makes sense if you have products to sell, like insulin or OxyContin. But if you want a blockbuster drug for Alzheimer's disease, and the astronomical profits that would come along with it, you need to have an actual product—an approved drug that is both safe and effective. Since that process begins with the basic research, and the industry is pulling up stakes, the firms have started outsourcing the discovery process and its inherent risks to others. The authors of the Pfizer/J&J study looked at the question of where the 81 percent of drugs that were not developed in-house came from. They found the following:

Some of them came to Pfizer and J&J from the acquisition of other pharmaceutical companies . . . Other . . . drugs originated in universities and academic centers. J&J's highest-selling product, infliximab (Remicade) . . . was synthesized by researchers at New York University in 1989 in collaboration with the biotechnology company Centocor. The original work [beginning in the early 1980s] showing its efficacy in rheumatoid arthritis was led by Marc Feldmann and Ravinder Maini at Imperial College London.

This is an oft-repeated scenario. Johnson & Johnson did not acquire the promising new product by harvesting the fruit of its own discovery process. It simply bought Centocor. Now known as Janssen Biotech, Centocor retains some of its original identity but only as a wholly owned subsidiary of J&J. This theme, with variations, occurs again and again throughout the industry. The discovery pathway begins in academia, transitions to a small-to-medium biotech firm, and ends with the small firm being acquired by one of the PhRMA giants.

This is not a romantic way to envision the drug discovery process, but it is remunerative. In the romantic version of the discovery process, an

Alexander Graham Bell or a Henry Ford or a Thomas Alva Edison functions as both inventor and leader of a company that bears his name. In the real-world version, the people who do the discovery work for a new drug make a bit of money but are largely forgotten in the PhRMA history books. Today, the entire reason for the existence of many a small biotech firm is simply to create a bit of intellectual property that can be monetized and sold to one of the big firms. It isn't a pretty story, but it works. We might even argue that it is theoretically a relatively efficient drug development process. But if this approach is so efficient, where is our cure for Alzheimer's disease?

THEY SAW THE WARNING SIGNS, AND YET THEY DID NOTHING

A successful cure for Alzheimer's disease would be worth billions. That's plenty of incentive to drive even the most cutthroat company to work quickly and efficiently to harvest a piece of that bountiful harvest. In theory, the profit motive should push companies to minimize costs, and since the bulk of the expenses of developing a new drug occur during the human trials, it seems a fair assumption that they would pick their clinical trial investments with great care. Yet over the past two decades the industry as a group has chalked up an unbroken string of costly Phase III trial failures (nearly 30 as of this writing). All of these failures have been trials of drugs designed to prevent or remove amyloid plaques from the brain.

During the heady days around the turn of the century and the extraordinary discovery that mice could be vaccinated against the amyloid peptide, it would have been financially and medically irresponsible not to push human trials to find out if the answer to Alzheimer's disease might be as simple as vaccination. The findings in mice were dramatic enough that even after the first human trial by Elan Pharmaceuticals had to be stopped in 2003, newly designed trials to minimize side effects were clearly warranted. But after a five-year follow-up of the Elan trial showed that the patients who were successfully immunized continued to decline at almost exactly the same rate as those who were not immunized, one might expect there to be some rethinking of corporate strategy. Fiscal responsibility would suggest a bit of caution in placing still more hundred million dollar bets on an anti-amyloid strategy. More preclinical work aimed at reexamining the underlying assumptions of the anti-amyloid approach would seem an appropriately cautious response. Diversifying the company's drug portfolio to put more resources toward testing compounds directed at other features of the disease would also seem to be a sound business strategy. To be fair, there were some efforts in these directions, but they paled in comparison to the public relations efforts to rationalize the failures and take one more shot on the amyloid goal.

Why? Why, despite a mountain of evidence that there were weaknesses in the amyloid cascade hypothesis did these financially successful companies never step on the brakes? They knew the data better than anyone. Hindsight is always 20/20, but if they had been more prudent with their expenditures, they would be billions of dollars richer today. There are no simple answers to the question of why an entire industry went so far off the road, but three explanations strike me as the most likely: stubbornness, greed, and bad advice.

Stubbornness is the easiest to understand. Titans of industry fashion themselves as smart and discriminating risk-takers. The caricatures of this attitude are the leaders of the oil and gas industry who have internalized the philosophy that when you go looking for oil, you're going to dig a few dry wells. If you're looking for oil, stubbornness can be seen as a virtue. When a well or two comes up empty, you sneer at those who turn tail and run. You trust your nerves of steel and your "instincts," and you go ahead and dig another well. "This time for sure." This attitude has a presence in the biotech field as well. Remember that 9 out of 10 compounds that start on the pathway to drug development have to be scrapped—the pharmaceutical equivalent of a dry well. But even the archetypal oilman knows that there are no oil or gas deposits in Massachusetts. Ambitious as they might be, after a couple of dry wells around Boston, they are not likely to keep digging. The value placed on high risk/high return ventures is baked into the psychology of the leadership of the PhRMA companies. It explains part of why the industry failed to heed the warning lights that were flashing red. The question is why they kept digging for oil in Boston.

Greed is a second, more complicated explanation. Viewed from outside of the industry, it seems evident that a savvy, greedy business leader would know when to stop throwing good money after bad. The loss of several hundred million dollars will certainly capture the attention of even the most aggressive CEOs and have them look for diversification and more biological exploration. This is long-term thinking, however. The greed that kept PhRMA in the amyloid-is-Alzheimer's race was all based on short-term thinking. I once asked a colleague in the industry about an anti-amyloid drug that had been totally lackluster in its performance in Phase II trials. Why, I wondered, had the company decided to launch a multicenter Phase III trial given how soft the Phase II data were. The answer I got was that the decision was a corporate one, not a scientific one. They had become vested in their product and realized that if they admitted failure in Phase II, they would pay a steep price on Wall Street. Besides, he said, maybe it will work in Phase III ("This time for sure").

His overall answer was exactly correct. Wall Street rewards good news, not good science. As a textbook example, consider the graph of the price of a share of Biogen over the year from December 2018 to December 2019. During the first six months of 2019, Biogen stock lost over 22 percent of its value. By the end of the year, however, much of that loss had been recouped (see figure 7.3). These downward and upward swings did not happen gradually over weeks or months; they happened over the course of just a few days. The sharp downturn came on the heels of the announcement that the trials of the anti-amyloid drug, aducanumab, were being halted because the trial oversight board had determined that

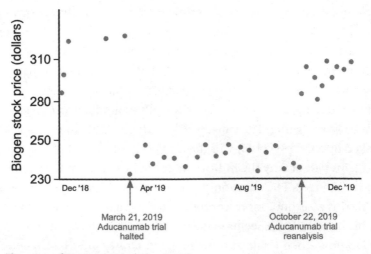

7.3   The price of Biogen stock during 2019.

continuing would be "fruitless." The upturn came over a few days in late October when, to the surprise of many in the field, the company announced that it had reexamined the trial data and, based on this reanalysis, had decided to file for FDA approval to market the drug. If you are a corporate leader, this graph says it all. Keep the good news coming, and don't ever let them smell your fear. Besides, even if the upturn is transient, there's a good chance you'll have moved on to a different company.

The final explanation for the sorry state in which the industry finds itself today is simply that they were getting bad advice. This is the saddest part of the story but also the one with the biggest impact. We've learned about the nature of this advice in the previous chapters. What began as maneuvers of academic politics—the double inflation of the meaning of Alzheimer's disease and the suppression of nonamyloid research—ended up creating a bevy of inbred experts who were repeatedly and exclusively called upon as advisors to the industry. So certain were these experts of their own models of Alzheimer's disease that they repeatedly reassured the CEOs they were on exactly the right course. If they would just start their trials a little earlier in the disease process, if they would just pick participants a bit more selectively, if they would just use a drug with a different way of targeting amyloid, they would be rewarded with the blockbuster drug they had been seeking. To the stubborn, greedy executives of the pharmaceutical industry, these words helped ease the pain of the long train of expensive failures. They offered hope and reinforced their self-image as smart and discriminating risk-takers. They were a life preserver tossed to a drowning person caught in very rough seas. They were also wrong.

The industry should have known better. It had the data, and it had statisticians to interpret them. It had the financial incentive to stop the hemorrhaging of money into the black hole of yet another Phase III trial. It had access to knowledge of the many alternative disease models that were out there. Smart money would have pulled back. There was no need for a Pfizer-like turning tail and running for the exits, but a calculated retrenching of resources would have been worthwhile. They could have worked on charting long-term plans to stop digging for oil in Massachusetts and start digging in Texas or Oklahoma. We are beginning to see the first glimmers of this pivot taking shape, but we have a long way to go. The experts that serve to this day as scientific advisors also should have known better. Their persistent recommendations to focus on an amyloid-only approach was rooted more in their own stubbornness than on truly thoughtful advice. Their promotion of their own ideas morphed into a need to suppress everyone else's thinking. They knew what the trial failures meant. They knew how weak the data were that drove the decisions to advance to Phase III trials. They owed the companies more objectivity in return for their consulting fees.

And they are still at it. The commentary surrounding the Biogen/Eisai decision to seek FDA approval for aducanumab is telling. Biogen's own interpretation of the data still leaves many questions, yet the experts are assuring us that "This time for sure":

We have had a terribly frustrating series of disappointments in the field. After the futility analysis of aducanumab and the multiple failures of BACE [beta-secretase] inhibitors, many were convinced we were barking up the wrong tree. I think these results, although complicated, should resurrect the enthusiasm for targeting amyloid.[6]

In 2019, this is magical thinking and frankly irresponsible.

The pharmaceutical industry is an indispensable part of the search for a cure for Alzheimer's disease. But while they are totally committed to the search, their reasons for joining the effort differ from ours as health care consumers. This divergence of interests—agency as it's known in the business world—has been strategically disastrous for the consumer. In the current business environment, the legally mandated mission of any large publicly traded pharmaceutical company (no matter what they say on their web page) is not to cure disease, or even to help people feel better. Their mission is to make money. Corporate behavior driven by any other motive is considered a dereliction of fiduciary responsibility to the company's shareholders. With this principle in mind, some of the puzzling decisions of PhRMA make a sort of perverse sense.

But we want a cure for Alzheimer's disease. And while the industry pursues its corporate interests, we are wasting time, we are wasting money, and we are wasting lives.

This is not how you study a human disease.

8

TESTING OUR MODELS: BREAKING BAD

The strong scientific rationale behind the development of the amyloid cascade hypothesis propelled it to a position of prominence in the field of Alzheimer's disease research. Even though other ideas were out there, they were largely suppressed. What we learned in chapter 5 was that they were suppressed not because they were inherently flawed. Each of these other models had its strengths as well as its weaknesses. Yet time after time, the strengths were ignored and model after model was bypassed because of the purported strength of the amyloid cascade hypothesis. This was a mistake.

One solution to any medical problem is almost never enough. As an example, think of the problems we would have today if we had stopped looking for antibiotics after Alexander Fleming found penicillin. In the modern era of multidrug-resistant "superbugs," we would be in big trouble. We are much better prepared for bacterial pandemics because we have erythromycin and vancomycin and tetracycline and bacitracin. They all kill bacteria. But each one does it differently, and having all of these arrows in our quiver offers us a much better chance of defending ourselves against microbial attack. The same is true for Alzheimer's disease. We can predict with confidence that some people won't respond to anti-amyloid treatment; some people will develop resistance to the treatment so that it stops working for them; some people will have intolerable side effects; and subtypes of Alzheimer's disease will exist that use different mechanisms to cause dementia. For these and a hundred other reasons we should never abandon promising alternative leads.

In the chapter 5 we dealt with the debating tactics that the field used to ensure the rejection of the many alternatives to the amyloid cascade hypothesis that had been proposed. What we didn't do was ask how good the amyloid cascade hypothesis itself was. As the objections that Hardy and Selkoe dealt with in their paper suggest, there are weaknesses in the theory. The question is how serious they are. To answer that, we have to do studies to test its predictions.

Two hypothetical tests come immediately to mind. The first would be to take a group of healthy people, put lots of amyloid in the brains of half of them, and keep the other half as a control. If the hypothesis is correct, then the people who get the amyloid should start the amyloid cascade and end up with Alzheimer's disease while the controls stay healthy. The second hypothetical test would be to take a group of people who already have Alzheimer's disease, take the plaques out of the brains of half of them, and leave the other half untreated to serve as a control. Removing the plaques should stop the cascade and therefore stop the progression of the disease. Sadly, because our brain's ability to repair itself after it is structurally damaged is almost zero, even stopping the cascade in its tracks wouldn't restore most of the lost functions. But if the cascade is stopped, we should expect the disease would not get any worse. This is disappointing news for a family whose loved one is severely affected by dementia, but for the family of a person in the early stages of Alzheimer's disease, keeping their loved one from getting worse would be nothing short of a medical miracle.

TESTING THE HYPOTHESIS I: ADD AMYLOID, CREATE ALZHEIMER'S DISEASE

Simple ethical considerations, not to mention a slew of international laws and treaties, prevent us from doing these exact experiments on human beings. But do not despair, nature has gone ahead and done some of the work for us. It turns out that if you randomly pick 100 healthy elderly people with normal cognition, you will find that 25–30 of them have amyloid plaques in their brain at a density high enough that a skilled pathologist might say that the person had Alzheimer's disease. The original discovery of amyloid plaques in the brains of healthy people came from microscopic experiments similar to the ones that Alzheimer performed on Auguste D. over 100 years ago. Researchers didn't expect to find anything when they looked for plaques in persons with no history of dementia. Imagine their surprise when they found that there is about a 30 percent chance that, with advanced age, the brains of cognitively healthy individuals are pretty well speckled with amyloid plaques. The same results were found using a newly developed dye known as Pittsburgh Compound B (PiB). PiB stains amyloid plaques and allows radiologists to visualize them in living people during a positron-emission tomography (PET) scan. Since its discovery in 2002, thousands of persons with normal cognition have been examined. The numbers are the same as those reported from the earlier microscopic studies. One-quarter to one-third of people with no clinical signs of dementia or even mild cognitive impairment (MCI) have enough plaques in their brain that they could be diagnosed as having Alzheimer's disease. How are we supposed to think about these results? The brain has the microscopic signs of Alzheimer's disease, but the person is just fine, thank you very much.

The plaque-positive cognitively intact people are an experiment of nature that allows us to perform the first of our two tests of the amyloid cascade hypothesis. Without having to inject it artificially, we have found a group of people who, for some reason, have ended up with lots of amyloid in their brains. What they haven't ended up with are significant memory problems or anything resembling Alzheimer's disease. It would appear that the amyloid cascade hypothesis has failed its first test.

Remember that a good hypothesis can fail one test and still provide useful information. So, we might want to gather a bit more data. For example, maybe these plaque-positive people were right on the verge of developing Alzheimer's disease. If we had just waited a little longer, they would have started to show signs of dementia. This seems pretty unlikely given the numbers involved. It's not as if 30 percent of the attendees at a 50th college class reunion (mostly 72-year-olds) show signs of Alzheimer's disease before the end of the year even though, if we were to do PET scans, 30 percent of them would have significant amyloid plaque burdens. Still, let's ask how long it is after people get plaques in their brain before they start showing signs of dementia. Is it weeks? Or maybe months? The answer turns out to be years.

Since we can check amyloid levels in a living person using PiB binding, we can compare how fast the 30 percent with amyloid plaques develop dementia. Let's stack the deck a bit and start with individuals that are already affected with MCI. When we do, we find that people with mild dementia and amyloid deposits in their brain move on to full dementia faster than those who have no amyloid.[1] But even with a head start on their amyloid deposits, after a full year only 20 percent of the amyloid-positive group go on to develop Alzheimer's disease. It would take three to four years for just half of these theoretically vulnerable people to progress to an Alzheimer's diagnosis. That seems pretty slow. If someone is already at the plaque stage of the amyloid cascade and their dementia is already starting to show, we might expect that things would move along a bit more quickly. It looks as if our brains are just not that much bothered by amyloid plaque build-up, at least on a moment-to-moment basis.

Our brains are not oblivious to amyloid, however, and here is where we should be glad that we hesitated to throw out the entire amyloid cascade hypothesis before we had made a few more observations. In this same experiment, the people who began with mild dementia but without amyloid plaques in their brain also progressed to Alzheimer's. They just progressed much slower. Only 7 percent of the PiB-negative (no amyloid) participants advanced to full dementia after a year. Six years into the study, well over half of these people had not yet started to show signs of Alzheimer's disease. The numbers of participants in this study is relatively small, but larger studies point to the same conclusion.[2] In general, having amyloid deposits in your brain increases your risk of developing Alzheimer's disease by about three- to fourfold over the next few years. That's informative, but remember a single copy of the APOE4 gene increases your risk even more. On the whole, it seems that not having plaques in your brain is a better way to live, but not that much better.

The full set of data argue that there is some role for amyloid in Alzheimer's disease, but they argue far more forcefully that amyloid alone is not a ruthless killer, at least not a lone assassin. Our brains can function perfectly well in its presence. As we get older, having amyloid in our brains seems to be a common part of our existence. This is distinctly counter to the predictions of the amyloid cascade hypothesis and should force us to consider major changes. It is hard to look at the time involved—years from the appearance of amyloid to the onset of clinical dementia—and still favor a model wherein the presence of amyloid starts a relentless cascade of nervous system death and destruction. There is no obvious biological reason that the steps of the cascade should take years to play themselves out. Plus, we still cannot be sure which way the causality goes. Amyloid tends to appear first, suggesting but not proving that it is the driving force. But we could still be dealing with a three-body problem of the type we spoke about in chapter 2. Indeed, given the long delay between the appearance of amyloid and the onset of dementia, a third force that independently leads to both the pathology (plaques and tangles) and the dementia is a plausible explanation for all of our data—one that we cannot disprove. For now, let's put such questions aside. We have more hypothesis testing to do. So far, it looks as though the amyloid cascade hypothesis is in need of modification, but we don't yet know how much. Let's see if we can test some of its predictions not in people but in mice.

EXPERIMENTAL MODELS OF HUMAN DISEASE

Human beings are terrible experimental models. Their genetics are messy, and their environment, diet, and life habits vary enormously. Unfortunately, it's the experimental preparation we're stuck with. If we are trying to cure Alzheimer's disease, there is no getting around the fact that we need to study it in humans. The question is "How?" Clinical trials are human experiments, but they are very expensive to run; each costs hundreds of millions of dollars and usually tests only one drug at one or two dosages. If, as is currently the case, we are missing big chunks of the biological picture of Alzheimer's disease, as a tool of discovery this approach is not going to be very efficient or very productive. To deal with this problem, researchers have identified a variety of creatures that are easily worked with in the laboratory yet still provide important information about human Alzheimer's disease. These other creatures are vastly cheaper to work with, and while they are not humans, studies with them have sparked many important discoveries. There are compromises to be sure. Practical issues of cost need to be balanced against theoretical issues of predictive value. This can be done, and over the years acceptable compromises have been reached in a wide range of organisms—monkeys, mice, fish, flies, and worms. Choosing among these models is both easy and hard. To begin with, there is no one best Alzheimer's disease model; the choice is made based on the question being asked. A quotation attributed to the British mathematician George Box very much applies here: "All models are wrong. Some models are useful."[3]

Studying the brain of a fruit fly might seem a ridiculous way to study a complex human disease. Flies don't get Alzheimer's disease, so it's wrong as a model. But a fly can remember things, and we can test its memory function in response to different drugs. What's more, it has an APP gene and a presenilin gene and we can manipulate its genetics to cause it to artificially deposit amyloid in its nervous system. This allows us to answer at least some basic questions about the mechanisms of memory loss in response to amyloid, even if we have to later prove that the same mechanisms apply in more complicated creatures such as you and me. The fly model is wrong, but it is very useful.

By far the favorite model of the Alzheimer's disease field, however, is the laboratory mouse. As a mammal, it shares much of our physiology, and its basic brain anatomy is similar. Looking at the two brains side by side, however (see figure 8.1), makes it plain that based on size alone, while it might be better than a fruit fly, as a model it's still wrong. But it has proven very, very useful. As with fruit flies, we can manipulate the mouse genome and create a genetic copy of familial Alzheimer's disease in a fellow mammal. The copy isn't perfect; overall, the regions of our human genome that code for proteins are only about 85 percent identical to those same regions in a mouse. But that's a lot closer than a fruit fly. For example, as we've already seen, these genetic copy mice produce amyloid plaques in their brains.

As written, the amyloid cascade hypothesis should apply to mice as well as to humans. The Hardy and Selkoe review, and many, many others, speak of the mouse's potential utility in preclinical Alzheimer's studies. Indeed, both the third and fourth watershed discoveries that we discussed in chapter 4 were made in mouse brain. By putting a familial Alzheimer's disease gene in the mouse genome, we can create mice that produce plaques in their brain. That memory problems came along with these plaques initially gave us real hope that these mice were a useful Alzheimer's disease model system. And, of course, being able to clear the plaques and correct

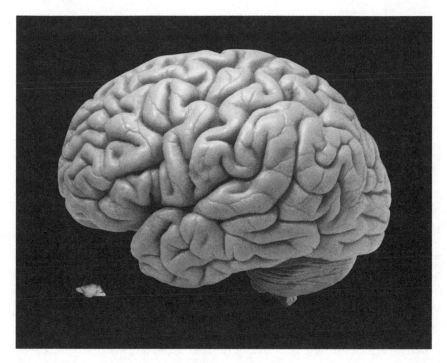

8.1 Relative sizes of the brains of mouse and human.

the memory problems left normally sober scientists jumping for joy. The joy was understandable. We all understood, as Box would remind us, that the model was wrong. And because of the way we went about studying them, they turned out to be not particularly useful either.

The condition of the mouse models in no way resembles a mouse version of Alzheimer's disease. The clinical pictures don't match up at all. True, the mice have plaque buildup early in their lives, but they have very few other health problems. They groom themselves and have no trouble eating and drinking. Watching them in their cage, it is difficult to tell them apart from their genetically normal companions. Contrast this with the appearance of humans in the Alzheimer's disease unit of any nursing home. Here there is no ambiguity about who the residents are and who the visitors are. The end stages of Alzheimer's disease come with near total immobility, an inability to care for personal needs, and difficulty eating. Even the most severely affected mice do not come close to this level of disability. So, the "clinical" picture of the mouse models is incomplete at best.

The picture in the brain is not much better. As exciting as it was to see plaques in the mouse brain when human APP was introduced, it slowly dawned on us that apart from these plaques most of the other structural changes found in the human Alzheimer's disease brain were missing. There were no tangles, there was no loss of nerve cells, and there wasn't even much loss of brain synapses (ping-squirt-ping). The brain picture was as incomplete as the clinical picture. The disconnect between mouse and human is seen in other areas as well. The familial (genetic) forms of Alzheimer's disease are caused by three known disease genes. *APP* is one; the other two are presenilin genes, *PSEN1* and *PSEN2*. Recall that prese-nilin is the active part of the γ-secretase. For familial Alzheimer's disease, mutations in the presenilin genes are four times as common as mutations in APP, so you would expect that the field would focus on mouse lines that carry presenilin gene mutations. Such mice were created, carrying a human presenilin gene instead of an *APP* gene, but to the great disap-pointment of the field no amyloid plaques appeared. The *PSEN* mice were pronounced as having no symptoms and were basically abandoned as models of Alzheimer's disease.

A mouse can never be a perfect model of Alzheimer's disease. Even so, if we look at the mouse data dispassionately, it has been useful, just not in the way that the proponents of the amyloid cascade hypothesis would have us believe. Since the first APP model mouse was reported, well over two dozen different Alzheimer's models have been produced. They have generated reams of good data, but the larger story that they tell has been almost entirely ignored. The story is that of an indepen-dent running of our first test of the amyloid cascade hypothesis. We've already seen that putting amyloid in the human brain does not lead to a relentless cascade of events that ends in profound dementia. What all of the mouse models tell us is that the same thing is true for mice. No signs of dementia appear in any model despite the fact that some of the animals live three-quarters of their two-and-a-half year lives with lots of amyloid in their brains. There are minor memory deficits, but when we run the second test of the hypothesis, we will see that even these are not at all Alzheimer's like.

TESTING THE HYPOTHESIS II: REMOVE AMYLOID, CURE ALZHEIMER'S DISEASE

The second test of the amyloid cascade hypothesis has also failed in both humans and mice. We can remove amyloid from the brain, and even though the amyloid cascade hypothesis predicts that we should stop Alzheimer's in its tracks, the data say otherwise. That we could do this second test at all was not certain. The assumptions of the day assured us that the plaques were more similar to concrete slabs than anything biological; nothing could dissolve them except strong acid. The dramatic plaque-free brains of the amyloid-vaccinated mice made it clear that this was a flawed view of the situation. Building on these unexpected but welcome vaccine results, a huge international effort was begun to test the effects of clearing plaques first in mice, then in humans—essentially a perfect run of Test No. 2.

Let's look closely at what happens to the mice who have had the plaques removed from their brains. The first reports were nearly unanimously positive. A variety of different vaccine strategies were tested, and each one led to reduced plaques and improved memory in the vaccine-treated animals. Not only did memory improve, so did all of the other behaviors that come along with having amyloid deposits in a mouse brain. In fact, the mental functions of the treated Alzheimer's mice returned to near normal. We need to pause here and ask what the mouse amyloid vaccine actually fixed. To ask that question, we need to revisit the mouse models and the criteria used for their adoption.

First, we should remember that, by using genetics as their tool, the architects of these new lines of mice were really only creating models of familial Alzheimer's disease. Modeling the sporadic form, 99 percent of Alzheimer's disease, was largely avoided. The field was confident in this approach first and foremost because of the dominance of the amyloid cascade hypothesis. According to this thinking, any way that you could get amyloid into the brain would be sufficient to trigger the cascade that led to Alzheimer's disease. Doing it with genetics would be the same, or so it was thought, as doing it some other way. Whatever we learned would apply to all forms of Alzheimer's. A second reason for confidence in the genetic-only models was that the experts assured us that there were no significant differences between the clinical and neuropathological symptoms of inherited and sporadic forms of the disease. That turns out to be only partly true. PiB imaging, in particular, has made it clear that the patterns of amyloid seen in familial Alzheimer's disease are quite different from those found in sporadic forms.[4]

The second important feature of the mouse models that we need to remember is that the original mice were created during the period of peak excitement surrounding the amyloid cascade hypothesis. Nearly everyone bought into the idea that too much amyloid drives Alzheimer's disease. There was very little questioning of the idea that "if you're not studying amyloid, you're not studying Alzheimer's." Thus, when the new mouse lines were rated for their success in mimicking the human disease, the main criterion used was "Do they get plaques?" As long as there were also a few memory problems, all was good. No need to methodically look for much beyond a test of spatial memory known as the Morris water maze. Other features such as apathy, depression, executive functioning, and loss of motor abilities were never part of the requirement of a good mouse model. Even the absence of tangles, the absence of neurodegeneration, the absence of a progressive worsening of the condition—all biological/chemical hallmarks of true Alzheimer's disease—were seen as mysteries to be studied later.

Looking back with 20 years of hindsight, it's pretty clear that what the gene jockeys had actually created were multiple models of brains with amyloid deposits. If you're a passionate believer in the amyloid-is-Alzheimer's view of life, that's enough. And when the production of plaques came with mild memory deficits and the vaccines cleared up both the plaques and the deficits, case closed. Mouse Alzheimer's disease has been cured. But if you have even a sliver of a doubt about the hypothesis, the amyloid mice are not a model of Alzheimer's disease; they're a model of a brain with plaques. Depending on your outlook, those are two different conditions that may or may not be related to each other. For those of us with slivers of doubt, while we applaud the removal of amyloid, we want details about the behavioral and memory changes that go along with that removal. In the end, those behavioral findings lead to serious concerns that not only are our mouse models wrong but it looks as if they have led us astray on everything except the plaques.

It's important to be cautious about how much a cure for mouse Alzheimer's might translate into a cure for human Alzheimer's disease. In chapter 1 we heard the story of Dorothy and saw how her mental function got relentlessly worse. This was happening because with each passing moment tiny bits of her brain were lost, as happens to everyone with Alzheimer's disease. By the end of a typical course of Alzheimer's disease, a quarter of the brain just disappears, and this cannot be fixed. By and large, once a function is lost, the loss is permanent. In the last year of Dorothy's life if there were a pill that would have allowed Dorothy to return to independent living, even if that included some mild memory or other neurological problems, I am quite sure that her daughter would have arranged for days and days of celebration.

It is in this context that the mouse models of Alzheimer's disease are most clearly defective. They do not mimic this progressive loss of function. Even though they are surely not normal when they are tested in the Morris water maze, they don't get worse. The implication is that the memory deficits in the mice, whatever their cause, must not be produced by the same process of deterioration found in Alzheimer's disease. And we can be 100 percent sure about this because of the results of the mouse vaccine trials. The memory deficits of the Alzheimer's mice are significant compared with their fellow mice without amyloid in their brains, but if we vaccinate the Alzheimer's mice, these deficits go away.[5] They are completely cured. One hundred percent normal. At first blush that's a wonderful finding, but a moment's reflection makes it clear that we have a problem. This is not at all the situation we would expect in human Alzheimer's disease. By the predictions of the amyloid cascade hypothesis, a true mouse model of Alzheimer's would have memory problems that started soon after the plaques appeared, and these problems would get worse and worse over time. An effective vaccine might stop the worsening, but if the model were an accurate mimic of Alzheimer's disease, even the best treatment could not restore function that had been lost.

Stopping the deterioration is the outcome measure used for nearly all human clinical trial designs. No one expects the scores on human memory tests to return to normal. In my own lab, we showed how this ought to look in the mouse. Instead of using memory, we used a property of individual neurons that are at risk of death in the Alzheimer's disease brain. These neurons try to divide, but they can't; instead of making two cells out of one, they get stuck in the middle of the process. We can identify the stuck neurons, and this led us to test whether an anti-inflammatory treatment of an Alzheimer's mouse model might have an impact. It did. We could stop more stuck neurons from appearing in the mouse brain. But once the neurons started their cell cycle, we could not make them turn around and go back to just being normal cells. This is what should have happened to the memory functions of the vaccinated mice. The vaccine should have done nothing in the short run. The effect should only have been seen in the long run. Treated mice should have remained the same while the untreated mice got worse. Instead, the untreated mice stayed the same, and the impaired memory function of the treated mice returned to completely normal.

Not only were the mice completely cured but the "cure" did not take months or even weeks; the memory functions were restored within days. This rapid response, while seemingly good news, is further evidence that the mouse memory deficits are transient and reversible. That means that they are at best distantly related to the degenerative processes that cause Alzheimer's disease. It also means that learning how to fix them will not guide us toward anything therapeutically useful in our fight against Alzheimer's.

A second feature of the vaccinated mouse response is still worse news for the amyloid cascade hypothesis. We've learned that vaccination not only will stop amyloid plaques from forming in the mouse brain but will eliminate plaques even after they've formed. We do not know exactly how long this takes, but it is likely weeks and probably months. Given this timeline, it is not surprising that when the brains of vaccine-cured mice are examined, a few days after the treatment, they have just as many plaques as untreated animals. Yet they are behaving normally. This is proof positive that the plaques themselves do not cause the memory problems. In fact, just as we suspected from the human data, they may not cause any problems at all. The mice behave normally, even when their brain is still full of plaques.

Let's put all these findings together. The field was sure that plaques were the cause of Alzheimer's disease. Because of this surety, mouse models were judged by their ability to make plaques. These plaques were associated with mild memory problems but nothing else resembling

Alzheimer's disease. Vaccinating against the Aβ peptide could remove the plaques, and, even before the plaques were gone, the behavior of the animals returned to normal. From the standpoint of the clinical symptoms of Alzheimer's disease, not only are our current mouse models wrong but we have made them nearly useless. Note that I did not say the models themselves were useless. My claim is that we have misused them, misinterpreted them, and destroyed the valuable lessons they were trying to teach us. If we had listened to their story, we would have had a treasure trove of data showing that amyloid plaques don't cause Alzheimer's disease. We would have taken a step back and realized that the mouse and the human data were in nearly complete agreement. Mice can have a head full of plaques and get only mild memory problems. Humans can have plaques in their brain and be mentally totally healthy. The data are trying to tell us that amyloid in the brain is not enough to cause Alzheimer's disease. We should have taken stock and accepted that the amyloid cascade hypothesis was in need of major revision if not outright rejection. Unfortunately, we did not listen to our own data. Instead we spent billions of dollars running human clinical trials based on data that were saying nearly the opposite of what we wanted to believe they were saying. Those expensive trials were all essentially the equivalent of running Test No. 2 in humans, and, as should have been predicted, they failed.

The human trials began with amyloid vaccines and, with respect to the intended purpose of clearing the plaques, the human trials replicated the mouse findings perfectly. The researchers running the very first vaccine trial, the one that had to be stopped because of the brain swelling, examined the brains of people under the microscope after they had died (of causes unrelated to the vaccination). When they did this, the extent of plaque clearance was impressive, perhaps even better than what had been seen in the mice. Later, more definitive evidence for this came with trials that looked at PiB binding after antibody treatment. The results were the same: active or passive immunotherapy can remove plaques from the human brain. Based on this observation, the amyloid cascade hypothesis would predict that the disease should stop getting worse. Cognition and all of the other neurological and psychiatric signs should remain about where they were when the disease started. Sadly, this second test of the hypothesis also failed.

A dispassionate reading of the results of the vaccine trials to date shows that they have been an unmitigated failure. Trial participants ended up no better for having had the vaccine. Worse, it turns out that vaccinating against amyloid turns out to be somewhat risky. The first active immunization trials are the clearest example of this. The Phase I trials for safety seemed successful. No adverse events were found, so Phase II trials were started almost immediately. This larger study produced some sobering results. Four of the participants experienced a massive encephalitis—a dangerous swelling of the brain—and nearly died. The trial was immediately called off, and the clinicians started pouring over the data to figure out what had happened. They realized that some of the modifications in the vaccination protocol that they had thought would be harmless turned out to be anything but. This led to two intelligent responses. The first was to change the protocols to avoid the problems encountered in the first run. The second was to continue to follow the participants in the trial after the danger of encephalitis had passed.

The changes to the protocol took many forms—different forms of amyloid, different delivery methods, and a different way of producing the antibodies. In the latter case, this meant using a technique called passive immunization. Active immunization is what happens when you get a flu shot. The shot contains dead virus or pieces of dead virus, and our immune system learns to recognize these harmless mimics. Should a real virus enter our body, our immune system "remembers" and instantly mounts a vigorous response. Passive immunization is what doctors do when they don't trust our bodies to do the work themselves. They engineer a custom antibody in the lab and deliver it directly into our systems. Instead of teaching our immune system what amyloid looks like so that it can rapidly respond to its presence, the doctors directly administer a prefabricated anti-amyloid antibody. There are two advantages to this approach. It provides antibody even to people whose immune systems can't learn to make it on their own, and it gets around the encephalitis problem because there is no need to stimulate an immune response. This lowers the risk that there will be an out-of-control response.

The revised methods did avoid the life-threatening overreaction of the immune system, but it turns out that removing amyloid plaques from the brain may not be a totally good thing. While the later, more sophisticated trials were encephalitis-free, brain scans showed that the participants receiving the vaccine still had a noticeable swelling of the brain. This response was given a benign-sounding name, amyloid-related imaging abnormality-edema, or ARIA-E. But naming a mild encephalitis after an operatic solo did not keep the trial review boards from insisting that it be listed as an "adverse event." The swelling seems to resolve after a time, but the long-term effects, if any, remain unknown.

By now hundreds of people have been successfully treated to remove the amyloid from their brains. These trial participants should have seen their disease progression stop. Unfortunately, as predicted by Test No. 1 where having amyloid in a normal brain didn't seem to interfere with cognition, removing it doesn't seem to make much difference either. For some trials the tracking of cognition has gone on for years. The follow-up studies have seen a few glimmers of change, but the significance of these is hotly debated. Optimism is a wonderful thing, but realism is a better long-term strategy. And realistically speaking, in none of the trials to date has there been any significant arrest of the normally expected cognitive decline.

TESTING THE HYPOTHESIS III: BLOCK AMYLOID FROM EVER FORMING

If using the body's immune system to remove amyloid doesn't work, perhaps we could take a different approach to testing the amyloid cascade hypothesis. We could think of that as Test No. 3: give a drug that prevents amyloid from forming in the first place. This approach has also been tried, and the results are even more disappointing than the results of Tests No. 1 and No. 2. Recall that the Aβ peptide is cut out of the larger APP membrane protein by two molecular scissors. The β-secretase cuts just outside of the membrane to free one end of the Aβ peptide. The γ-secretase cuts within the membrane to free the other end of Aβ. Drugs have been made that inhibit both of these secretases. Blocking either one of them should totally eliminate the body's ability to make the Aβ peptide. Without Aβ there are no oligomers, no plaques, and theoretically no Alzheimer's disease. Both types of drug have been tested in humans, and in both cases the drugs worked as designed at the level of the biochemistry. Both blocked the appearance of Aβ. Unfortunately, when given to humans, they both led to serious side effects. These adverse events were bad enough that all of the secretase trials were halted before they reached their planned endpoint. In the case of the β-secretase inhibitors, the effects were worrisome but appeared to stop when participants were taken off the drug. The trial nonetheless was halted because the data oversight board, based on a preliminary analysis of the cognitive status of those treated with drug, proclaimed that there was "virtually no chance of finding a positive effect."[6] Not even a glimmer of hope.

The situation with the γ-secretase inhibitor (semagacestat) was beyond disappointing. Trial participants on the drug not only had side effects but their mental performance actually got worse. I was able to attend a workshop where the results of the semagacestat trials were described. Despite the recent trial failure, the presenter wanted to put a positive spin on the findings and offered the audience the observation that the results were truly noteworthy and, in many ways, a huge success. The audience was assured that this was actually the first time that any Alzheimer's disease drug had significantly altered cognitive performance. I thought to myself, "This has got to be the slickest spin doctor I have ever met. The drug altered cognition all right . . . *in the wrong direction*. It made people worse! How can you call this a success?" So, during the question period after the talk, I asked what I thought was an obvious question. "If we know that we can move cognition by altering γ-secretase activity, and if inhibiting it makes cognition worse, shouldn't we be looking for drugs that *increase* its activity?" This simple question was met with a profound silence in the room. The logic of my question was clear, and the data supported that approach. There was nothing but silence because increasing γ-secretase was the exact the opposite of what the amyloid cascade hypothesis would predict should be done. That's because increasing γ-secretase activity should increase the amount of Aβ that could be made. Since no one in the room was prepared to even think about abandoning the amyloid hypothesis, the only possible response was silence.

Though little discussed, these findings in humans were precisely predicted by earlier work in the mouse. In the mouse, use of genetics or pharmacology to inhibit γ-secretase leads to sick mice. If you genetically eliminate γ-secretase in the entire mouse, development stops when the mice are tiny embryos. If you inhibit it just in neurons and just in the adult, there is significant death of nerve cells. There is no reason to doubt that these toxic consequences of losing γ-secretase activity were exactly why the semagacestat trials failed.

GRADING THE TESTS: THE AMYLOID CASCADE
HYPOTHESIS FLUNKS THE COURSE

Skilled debaters are evaluated on how capable they are of mounting an intelligent and vigorous promotion of their position. Deciding which team wins a debate is less about which side of the argument has the most merit and more about which side of the argument has the cleverest team. Resolved: smoking should be banned because it is a public health hazard. We could set up a debate on that resolution, and even today lawyers for the tobacco industry could construct a clever case opposing it. But biomedical science is not a debate. New drugs and new treatments need to be designed on the basis of what is in the best interest of our collective health, not on the basis of who has the best lawyers. We must listen to our data and take action accordingly. There will always be counterarguments; certainty is a luxury we are not entitled to. Remember that a hypothesis can never be proven to be 100 percent true; it can only be tested and tested and tested until we find the limits of its applicability. There is no winning or losing; there is only more or less confidence in an idea's validity.

After considering the evidence, I would argue[7] that the amyloid cascade hypothesis has proven itself to be so severely limited that it is virtually worthless. It fails the three basic tests we applied:

Test No. 1: In both humans and in mice, adding amyloid to healthy brains does not start the amyloid cascade.

Test No. 2: In humans, taking amyloid out of the brains of people with Alzheimer's disease does not stop the disease.

Test No. 3: Blocking the formation of amyloid from the APP precursor does not stop the disease and actually makes both humans and mice sicker.

Based on these failures, we should reject the amyloid cascade hypothesis. The Aβ peptide does not start a cascade that leads to Alzheimer's disease. This is not an argument against any role for amyloid. The hypothesis has failed its tests and flunked the course, but that does not mean that amyloid is good for us. People with amyloid deposits are more likely to develop Alzheimer's disease; mice with amyloid have deficits in spatial memory. But the increased risk in humans is not great, and the mouse deficits are mild and bear almost no relationship to the situation in human Alzheimer's disease. The responsible course of action therefore is to propose a new hypothesis to explain the causes of Alzheimer's disease.

HOUSTON, WE HAVE A PROBLEM

In chapter 12 I will offer one possible new hypothesis. But first I have to ask you to consider a rather uncomfortable implication that has been building up over the past few chapters. If the amyloid cascade hypothesis is rejected and we agree that amyloid is not the cause of Alzheimer's disease, then what in the end is Alzheimer's disease? In chapter 1, I told you that nearly every expert I spoke with had a somewhat different definition. After we heard the story of Dorothy, I offered a working definition:

Alzheimer's disease is a late-life disease that, over the course of many years, destroys normal brain function in a progressive and irreversible fashion. Throughout the advance of the disease, the person is largely unaware of the dramatic nature of the changes that are happening to him or her. The inability to form new memories is one of the first hints that there is a problem. The person then begins to lose his or her ability to perform complex tasks. As the disease progresses, language skills and reasoning deteriorate as does the ability to make judgments. Personality changes such as depression and apathy set in along with unexpected emotional outbursts, aggression, and agitation. The ability to navigate worsens, leading to helpless wandering. For each of these changes, the severity of the dysfunction increases with time. Through most of the disease process the physical health of the person with Alzheimer's remains strong despite the continuing mental deterioration. By the end stages, however, the person becomes bedridden, incontinent, nonverbal, and nonresponsive.

Rereading that definition now, I trust you will see how it doesn't describe the disease process; it only describes the symptoms. This is fine for a family member to describe the situation that a loved one finds themselves in, but this is not a useful starting point for a biologist whose goal is to find a cure. What's going on with the cells of the body and the molecules inside these cells, and how can we stop the bad behavior of cells and molecules? To answer those questions, we are going to need a biological definition of Alzheimer's disease. By the end of the next chapter I hope to convince you that we don't have one. And not having a definition to guide us is certainly no way to study a human disease.

9

WHAT IS ALZHEIMER'S DISEASE?

How odd that we are now two-thirds of the way through a book on Alzheimer's disease and we have to reask the question "What is the definition of Alzheimer's disease?" The message that I hope has come through in the past eight chapters is that the definition has been a moving target ever since the term was first coined. As a result, over the past 100-plus years, what started as a case study has ballooned into a label for the majority of all age-related dementia.

Inflating the definition of "Alzheimer's disease" could have been harmless. Over the years if the field had come up with cures, all of this posturing and political maneuvering would have been nothing but a historical footnote in a larger success story. Unfortunately, nothing like that happened. Kraepelin wanted to push the idea that the abnormal deposits were the cause of Auguste D.'s dementia. In doing so, he used the plaques and tangles as the basis for applying the name "Alzheimer's disease" to any case of presenile dementia in which these deposits could be found. This relatively small inflation was followed by the one we learned about in chapter 6 that was promoted by the NIA and the Alzheimer's Association. This second much larger inflation was an attempt to apply the label "Alzheimer's disease" to as much of late-life dementia as possible. The goal, once again, was strategic, but in this case, it was not even remotely related to a philosophy of brain function; it was all about getting a bigger slice of the NIH pie to the young National Institute on Aging.

The problem caused by these two inflations was the distortion of definition. For Alzheimer and Kraepelin, the rare form of early-onset dementia they named Alzheimer's disease was, they believed, caused by the deposits they had seen in the brain of Auguste D. That first linkage of deposits and dementia was the Trojan horse that released the soldiers of the second inflation. If we were going to call most late-onset dementia "Alzheimer's disease," then we had to accept Kraepelin's definition of Alzheimer's as being defined by deposits. We were forced into declaring that the majority of late-life dementia was caused by amyloid. Instead of listening to our data, we've essentially redefined the disease to fit our preconceived ideas. How did we get ourselves into a situation where the presence of a brain deposit that is at best loosely related to any specific disease is required to define a person as having Alzheimer's disease? To answer that question, we need to return to the history of the second inflation.

The article most commonly cited as the manifesto of this effort was written by Robert Katzman and bore the fearsome title "Editorial: The Prevalence and Malignancy of Alzheimer Disease: A Major Killer."[1] Katzman began his two-page editorial by arguing that there was no really significant difference between the relatively rare condition that was known at the time as Alzheimer's disease and the far more common condition known as senile dementia. He then went on to argue that dementia was badly underdiagnosed as a cause of death. The death certificate might list pneumonia, but that ignored the fact that without the disability of dementia, the pneumonia would never have happened. He estimated that if the cause of death were adjusted to honestly reflect this fact, dementia was arguably "a major killer." The real purpose of the editorial, however, was to argue for equating senile dementia with Alzheimer's disease. This was a bit of a stretch. As we have learned, Alzheimer's disease was recognized only as a rare early-onset form of *pre*senile dementia. The more common senile dementia was seen as simply a consequence of aging.

To bolster his case for their equivalence, Katzman cited earlier clinical speculation that Alzheimer's disease and senile dementia were similar in their symptoms. He also cited a pair of papers published a few years earlier.[2, 3] The authors of these earlier works, Tomlinson, Blessed, and Roth, examined the autopsied brains of 50 people who had died with a diagnosis of dementia and 28 people who were judged to be dementia-free.

Katzman wrote, based on his analysis of these two papers, that when comparing the microscopic appearance of the brain of a person who had died with Alzheimer's disease with one who had died with the more common senile dementia, "The pathological findings are identical—atrophy of the brain, marked loss of neurons, neurofibrillary tangles, granulovacuolar changes, and neuritic (senile) plaques."
The problem is that this is not exactly what the two papers had claimed.

The two works by Tomlinson, Blessed, and Roth were major accomplishments. They invested a great deal of time and patience to carefully score many brain regions of many different autopsy cases for the presence of plaques, tangles, blood vessel changes, and other abnormal features (granulovacuolar degeneration—GVD). Inconveniently for Katzman's argument of identity, however, was the fact that the authors had specifically noted that of their 28 controls, "Many brains showed . . . senile plaques, Alzheimer's neurofibrillary change [tangles] and granulovacuolar degeneration." As for their 50 cases of dementia, only 70 percent of them had more deposits than the controls. Five cases (10 percent) could not be diagnosed with confidence by the microscopic appearance of the brain, and 40 percent had changes that would have led to a non-Alzheimer's diagnosis. In the end, only "50% were considered to be cases of senile dementia showing the histological features of Alzheimer's disease." This is hardly a rock on which to build the claim that senile dementia and Alzheimer's disease are one and the same.

The Tomlinson papers are important touchstones in the field, but the question for us is how to think about their data. On the one hand, it's clear that serious plaque deposits are only found in people with dementia; healthy controls don't have as many. On the other hand, only half of the people with dementia have a lot of plaques in their brains. Nearly a fifth had few or none. Plus, there are the people with normal cognition but quite a few plaques in their brain. The pie charts shown in figure 9.1 illustrate the data. The darkest wedges indicate cases that had lots of plaques, and you only see this situation in the group with dementia. But the dementia pie also has light gray wedges, meaning no serious Alzheimer's-like pathology. The reverse is true for the controls. The light gray wedges account for 70 percent of the control pie, but there is that darker gray wedge (lots of plaques) that accounts for 29 percent

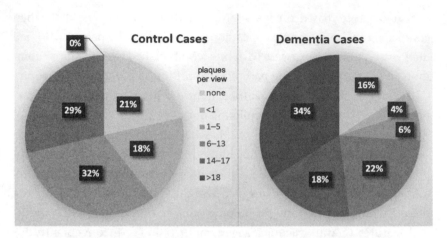

9.1 Plaque densities in the brains of individuals with and without dementia.
Source: Data are from B. E. Tomlinson, G. Blessed, and M. Roth, "Observations on the Brains of Non-demented Old People," *Journal of the Neurological Sciences* 7 (1968): 331–356; B. E. Tomlinson, G. Blessed, and M. Roth, "Observations on the Brains of Demented Old People," *Journal of the Neurological Sciences* 11 (1970): 205–242.

of the control pie. What Tomlinson and his colleagues realized was the following:

No qualitative histological feature . . . was encountered in the dements that was not found in the controls; only the degree to which the change occurred, or the distribution of the change differed in the 2 groups.

Thus, Tomlinson, Blessed, and Roth argued that Alzheimer's-like microscopic changes were strongly correlated with dementia. They are clear that in reaching a final diagnosis they considered Alzheimer's-like microscopic changes to include tangles, GVD, and loss of neurons in addition to the number of plaques.

Think of this conclusion using the language of our diagram from chapter 2 (figure 2.1). What Tomlinson, Blessed, and Roth did was to add two new shapes to our diagram—a diamond and an oval—but they took away all of the arrows (see figure 9.2). They were not arguing cause and effect. They were arguing that if you found the four shapes together (meaning, inside the box of correlation), you could call that Alzheimer's disease. Their main point was that most late-life dementia had these pathologies coexisting. They urged that their common co-occurrence argued for calling dementia with these changes by a single name. They explicitly did

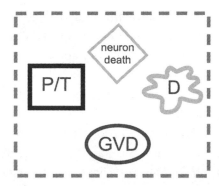

9.2 Correlation, but not causality. All shapes within the large dashed-line box are cor-related. No arrows between the shapes indicates that there are no paths of causality from one shape to another. P/T, plaques/tangles; D, dementia; GVD, granulovacuolar degeneration.

Sources: In the findings of B. E. Tomlinson, G. Blessed, and M. Roth, "Observations on the Brains of Non-demented Old People," *Journal of the Neurological Sciences* 7 (1968): 331–356; B. E. Tomlinson, G. Blessed, and M. Roth, "Observations on the Brains of Demented Old People," *Journal of the Neurological Sciences* 11 (1970): 205–242.

not single out plaques as the driving feature of this nomenclature. Also, they were silent on how the lines of causality were drawn (the arrows in figure 2.1). In other words, they took no stand on the underlying biological mechanisms that lead to Alzheimer's disease.

Katzman's argument engages in a bit of sleight of hand. He claims first that senile dementia really is Alzheimer's disease, just with a later age of onset. Then, since dementia is a major killer and dementia is really Alzheimer's disease, Alzheimer's must be a major killer. But Tomlinson and colleagues had been crystal clear that only about half of the cases of senile dementia actually have the pathological appearance of Alzheimer's disease. It also bears mentioning that while plaques are more common in someone with a diagnosis of dementia, they are not unique to Alzheimer's. People with Parkinson's disease, Huntington's disease, traumatic brain injury and epilepsy also have amyloid plaque deposits. Even children can have them. Years after Katzman's editorial, Heiko Braak, a well-respected German pathologist, wrote that when he looked at a large sample of autopsies that had been done on brains of all ages, only 10 out of 2,332 were completely free of plaques and tangles. Why would we define a disease by something that is so nonspecific?

Katzman wasn't wrong about the idea that dementia kills no matter what it says on the death certificate. Speaking to the "malignancy of Alzheimer's" was strongly worded but appropriate. In the struggle to define Alzheimer's disease, however, his proposed equivalence between presenile and senile dementia was deceptive. First of all, half of the cases of senile dementia don't have plaques and tangles, which according to everyone from Kraepelin to Tomlinson, Blessed, and Roth means they don't have Alzheimer's disease. Second, the boundaries between Alzheimer's and non-Alzheimer's dementia are fuzzy. As we saw in the pie chart in figure 9.1, deciding whether or not someone has Alzheimer's is based not on *whether* they have plaques but rather on how many they have. The amount of "plaque-iness" needed to diagnose Alzheimer's disease is arbitrary at best and debated to this day. The intent of Katzman's argument, for reasons we discussed in chapter 6, was to expand the definition of Alzheimer's disease to help get the NIA started. In this effort, it succeeded beautifully. The unintended negative consequence of this success, however, was to permanently tie amyloid plaques to the definition of Alzheimer's disease.

Katzman's inflated view of Alzheimer's disease soon took hold. By 1980, it had earned a place in the third edition of the *Diagnostic and Statistical Manual of Mental Disorders (DSM-III)*. Subsequently, attempts were made to precisely define the pathology needed for a diagnosis of Alzheimer's disease.[4] Eventually, there were so many "authoritative" sources on how to diagnose Alzheimer's disease that a well-meaning clinician could surely have been forgiven for getting frustrated about what this newly inflated condition called Alzheimer's disease really was and whether it applied to the elderly person sitting in his or her office.

To attempt to deal with this, a working group was assembled under the auspices of the NIA and the Alzheimer's Association (then known as the Alzheimer's Disease and Related Disorders Association—ADRDA) to formalize the clinical diagnosis of Alzheimer's disease.[5] The working group came up with a list of criteria to diagnose what they called "PROBABLE Alzheimer's Disease [all caps is their emphasis]." The list they came up with was similar to the definition we established in chapter 1 in that it was based only on the clinical picture presented by the patient:

Criteria for a diagnosis of Alzheimer's disease

- Dementia established by clinical examination
- Deficits in two or more areas of cognition (problem solving, language, attention, etc.)
- Progressive worsening
- No disturbance of consciousness
- Onset between ages 40 and 90, most often after age 65

But the group went one step further. They established criteria for a diagnosis of "DEFINITE" Alzheimer's disease. For this score, you first had to have a clinical diagnosis of dementia (the five criteria above), but you also needed "histopathologic [i.e., microscopic] evidence obtained from a biopsy or autopsy." And what did they consider definitive histopathologic evidence? Remarkably, they didn't say. They don't even discuss the issue. I have the feeling that this was because the experts couldn't agree on an exact set of criteria. Our well-meaning clinician had to wait until the following year when a separate summary of the workshop was published.[6] Buried in this document we learn that the pathologist has to do a lot of plaque counting to reach a conclusion. The counting criteria are listed here:

Khachaturian pathological criteria for Alzheimer's disease

- In people less than 50 years of age, there must be more than 2–5 plaques per microscopic field
- 50–65 years of age, there must be more than 8 plaques per microscopic field
- 66–75 years of age, there must be more than 10 plaques per field

Note that this workgroup was arguing that if you're younger, you need fewer plaques for a diagnosis of DEFINITE Alzheimer's disease. This is actually at odds with Katzman's assertion that there was no meaningful difference between the historical definition of Alzheimer's disease (presenile dementia) and senile dementia. It also undercuts the Tomlinson et al. papers whose subjects were mostly in their mid-70s.

The details are important to the aficionado, but it's the big picture that is important to us. Alzheimer's disease was to be defined by the presence of plaques—a feature that is not present in 15 percent of the people with a clinical diagnosis of Alzheimer's, and a feature that is present in people of all ages including 30 percent of elderly people without any cognitive impairment. Tangles are ranked as "helpful" but not necessary; GVDs are not even mentioned, nor is neuronal death. If this doesn't make sense to you, it's because it doesn't make sense.

The field could not be bothered with such details, however. In 1986, building on the momentum of the working group, the Consortium to Establish a Registry for Alzheimer's Disease (CERAD) was established by the NIA. The sponsor alone should tell you something. Handpicked clinicians and experts were shown videos of typical cases of Alzheimer's disease to make sure everyone was on the same page when they decided on a clinical diagnosis. But the final word—the gold standard as it was called— was the neuropathological presence of amyloid plaques. If there were not enough plaques, whatever the neurologist had found was not Alzheimer's disease. The CERAD community was meticulous in its attempts to make the clinical evaluations of the entire Alzheimer's medical community as consistent as possible. But despite yearly meetings and rigorous practice sessions, complete with review of videotapes, the pathologists pronounced the neurologists to be in error almost 15 percent of the time. Thus, unlike most other diagnoses in the *DSM*, when it came to Alzheimer's disease, the CERAD group put the imprimatur of the federal government on the notion that the presence of plaques, not the clinical presentation, was the gold standard of Alzheimer's disease diagnosis.

So, who died and made the pathologist king? There was really no good reason for pathology to trump neurology. You can get a definitive diagnosis of any number of complex brain diseases—autism, depression, schizophrenia, epilepsy, and many others—without any pathological study or live imaging of the brain. If a child psychiatrist diagnoses a young boy as having autism, there is no need for a brain scan to test the psychiatrist's skill. For the purposes of treatment, the child has autism. If a neurologist diagnoses a person as having Parkinson's, that's the diagnosis. They don't wait with bated breath to find out whether there were α-synuclein deposits in the brain. For the purposes of treatment, the person has Parkinson's disease. Late-life diseases like Parkinson's and Huntington's do have characteristic brain abnormalities, but it is the presentation of the clinical symptoms that allows physicians to have confidence in their diagnosis. To be fair, if an autopsy is done and there are no deposits, the diagnosis is questioned, but not rejected.[7] The clinical diagnosis overrules the pathology. As it should.

A bit of historical context is helpful in understanding why Alzheimer's disease was somewhat unique in crowning the pathologist as king in the area of diagnosis. Since the beginning of human history, healers had been able to diagnose a disease, but treating it was an empirical art more than a rigorous scientific enterprise. Aspirin for pain and quinine for malaria were available because somebody had noticed that administering these compounds helped. Occasionally a chemical change made them work better, as with aspirin, but still, no one knew how they worked at the level of cells and molecules. They just did. Beginning in the mid-twentieth century, however, our deepening insight into human biology held out the promise of *evidence-based medicine*—a new concept that we could design drugs and treatments based on our knowledge of biology and chemistry. Led by the excitement surrounding the discovery of the structure of DNA and the triumph of the Human Genome Project, there was great optimism that we could describe how disease worked at the atomic level.

This was the intellectual environment in which the inflationary ideas of Katzman, Terry, Khachaturian, Butler, and others were taking root. Linking biochemistry, in this case amyloid deposits, to disease was cutting-edge thinking for its time. The lionization of Alois Alzheimer for his assertion that plaques caused dementia can be seen in some ways as an attempt to keep up with the geneticists. It was a reaction to the growing pressure in biomedical research circles not just to diagnose disease but to uncover its biological basis. No brain scientist wanted to be left behind while the molecular biologists stole the show. No one wanted to be seen as old-fashioned by looking for cures empirically (no more aspirin or quinine, thank you very much). The field desperately wanted to find the brain equivalent of the Human Genome Project to wow the world. Solving Alzheimer's disease would certainly be a big wow.

It was the intersection of this history with the inflationary push in the definition of Alzheimer's disease that presented a unique problem to the field. The leaders were betting that the biochemistry of amyloid would be the launchpad for a moon shot that would bring them fame and glory. This had an outsized effect on the Alzheimer's field because the idea of a moon shot amplified the pressure to inflate the definition of Alzheimer's disease. The reasoning was the same as that used to try to increase NIA's funding. The field needed a fearsome challenge. Curing the original narrowly defined "Alzheimer's disease," a rare form of presenile dementia, was more like putting a human in low-earth orbit. It would be far more exciting to cure a condition with the "prevalence and malignancy of Alzheimer disease." For real fame and glory, we needed to identify our enemy as "a major killer." We were on our way to the moon.

DOUBLING DOWN

By including neuropathology as the gold standard of diagnosis, the working group all but made permanent the connection between Alzheimer's disease and plaques. There were caveats, of course, and the admission that additional studies were needed. Predictably, however, for the majority of the field, those caveats were basically ignored. In the years following there were tweaks to the definition so that tangles as well as plaques might be considered,[8] but there was no real concerted effort to reexamine our definitions. Then our clinical trials started failing, our antibody trials in particular.[9] In the basic research laboratories of the world, data kept accumulating that violated the expectations of an amyloid-only definition for Alzheimer's disease biology. In response the NIA began to recognize that there was a "broad consensus . . . that the criteria should be revised to incorporate state-of-the-art scientific knowledge."[10] The result was a truly comprehensive review—a compendium of four papers comprised of reports of three working groups plus an introductory summary. Together, these publications correctly recognized a long overdue need to revisit how we define Alzheimer's disease. Coming as it did, a full quarter century after the CERAD working group met, it would have been an ideal platform to announce the decision to cut the definition of Alzheimer's disease loose from the presence of plaques. Instead the experts in the field doubled down and bet the store on amyloid. In doing so, they made the entire situation much, much worse.

I had been asked to critique an early version of this report and sent in comments that were later published citing both strengths and weaknesses.[11] I was confident that the authors would take my critique into account before they went public with their ideas. Silly me. I still remember the first public announcement of the new "diagnostic guidelines" in

July 2010 in Honolulu at an international Alzheimer's meeting. I sat in the audience as the experts from the working group rolled out the summary of what they had come up with. By the time they were halfway through the presentation, I realized that they hadn't changed a thing. They had made the definition of Alzheimer's disease all about amyloid. How, I wondered, could they possibly justify increasing rather than decreasing the field's obsession with Aβ? Had they slept through the last 15 years?

The document they produced was entitled "The Recommendations from the National Institute on Aging-Alzheimer's Association Workgroups on Diagnostic Guidelines for Alzheimer's Disease." Let's dissect its contents by beginning with a quote from the introduction.[12] The authors were reviewing the history of how the NINCDS-ADRDA (CERAD) criteria for diagnosing Alzheimer's disease evolved in the way that they did:

The original NINCDS–ADRDA criteria . . . were designed with the expectation that . . . AD, like many other brain diseases, always exhibited a close correspondence between clinical symptoms and the underlying pathology, such that (1) AD pathology and clinical symptoms were synonymous and (2) individuals either had fully developed AD pathology, in which case they were demented, or they were free of AD pathology, in which case they were not demented (at least not because of AD).

This was a very skewed reading of history. The CERAD publications were very clear that more work needed to be done linking pathology to disease. That earlier workgroup flirted with the idea of a close correspondence, but "synonymous" was certainly a most unfortunate word choice. Synonymous is defined as "having the same meaning." Essentially, then, this 2010 workgroup was defining Alzheimer's apples—what our grocery store analogy would call red apples with worms in them. It is indeed a definition, but it has no basis in biology and hence has little value.

The three reports from the working groups describe how to diagnose Alzheimer's disease in the clinic, how to diagnose an early stage of Alzheimer's known as MCI, and how to think about healthy people who have plaques in their brain. You might ask, "Where's the pathology paper?" That would be a great question since all three papers cite pathology in establishing this new definition of Alzheimer's disease. However, just as it was 27 years earlier with CERAD, we all had to wait a full year before the fifth and final paper in the series was published.[13]

Writing the first paper was a clear struggle for its authors. Their task was to come up with a clinical definition of Alzheimer's disease, but they were hamstrung by the pathology gold standard. In the end, they decided to separate clinical and pathological Alzheimer's disease. They defined "all cause" dementia and added several useful refinements to the CERAD criteria. Then they ran into trouble because they had to acknowledge that "probable" and "possible [less certain]" Alzheimer's disease can both occur with or without "evidence of the AD pathophysiological process," meaning evidence of amyloid deposits. This amyloid evidence is called a biomarker because it is a biological marker for the disease—a surrogate. To appreciate the struggle this group had, it helps to quote them directly:

In persons who meet the core clinical Criteria for probable AD dementia biomarker evidence may increase the certainty that the basis of the clinical dementia syndrome is the AD pathophysiological process. However, we do not advocate the use of AD biomarker tests for routine diagnostic purposes at the present time. There are several reasons for this limitation: (1) the core clinical criteria provide very good diagnostic accuracy and utility in most patients; (2) more research needs to be done to ensure that criteria that include the use of biomarkers have been appropriately designed, (3) there is limited standardization of biomarkers from one locale to another, and (4) access to biomarkers is limited to varying degrees in community settings.

This one paragraph highlights the debates that the working group must have been having with its members, and it offers us a peek at the controversies in the field at the time. The task before the group was to come up with a recommendation for how practicing physicians should decide whether or not a living person, sitting in front of them in their office, had Alzheimer's disease. This was a laudable and practical goal, and in laying out their new recommendations for diagnosis, they were thoughtful and quite detailed. But reading their words, it is quite clear they were not about to be drawn into saying in print that amyloid, or any other "biomarker" (like tau), should be used to define Alzheimer's disease. They made a clean and explicit separation between a clinical diagnosis of Alzheimer's disease dementia and what they called the "pathophysiological process of Alzheimer's disease." What the group essentially said was that in the clinic the presence or absence of amyloid is just a piece of information that can be helpful in reaching a diagnosis, and nothing more. Take that, Dr. Gold Standard Pathologist.

There is no way to read this paper as anything other than a yellow warning flag being furiously waved on the amyloid-is-Alzheimer's racetrack. Look at the four objections against the use of biomarkers listed in the quote above. Translating from the original doctor speak, the first objection says clinicians have a pretty good track record when it comes to diagnosing Alzheimer's. We don't need no stinking amyloid, thank you very much. The second says, you pathologists and biochemists talk a good game, but you don't have enough data to make a solid case for us using your biomarkers. The third says, you folks can't even agree among yourselves how to use the biomarkers. The fourth says, let's be practical. Even if you had more data, a typical primary care physician has neither the time nor the training to meaningfully use biomarker data in their office. What the authors are powerfully arguing is that Alzheimer's disease is defined by its clinical symptoms. The amyloid cascade hypothesis offers one view of a biological cause of these symptoms, but as of 2011 there simply is not enough data to accept or reject it.

I agreed then, and now, nearly 10 years later, I still agree.

The second paper in the series[14] was important for the attempt of the working group to define what is known as MCI (mild cognitive impairment). In 2011 this was a relatively new area for the field and reflected a growing certainty that there was an early stage of Alzheimer's disease dementia that could be separately defined and studied. The goal was a useful one: to try to clinically identify Alzheimer's disease as early as possible so that treatment could begin when the probability for meaningful impact was the greatest. These authors were a separate group of neurologists. They too wrestled with how to incorporate biomarkers into their recommendations. In the end, they conclude, "Considerable work is needed to validate the criteria that use biomarkers and to standardize biomarker analysis for use in community settings." Like the clinicians in the first paper, they are willing to say that people who have no evidence of amyloid or tau are "unlikely" to have MCI due to Alzheimer's disease, but they add the caveat that ". . . such individuals may still have AD, . . . [but for these patients] . . . a search for an alternate cause of the MCI syndrome is warranted." As with the first group, the MCI paper is arguing that while evidence of amyloid and tau may be useful information, it is not definitive.

Together, the first two papers in this well-cited series can be read as struggles to keep the clinical picture in the forefront of our definition of Alzheimer's disease. You will recall, from back in chapter 1, that this was the approach we took in our first attempt to answer the question "What is Alzheimer's disease?" For us as friends and family members it is the clinical symptoms that matter to us the most. These are the symptoms for which we are most anxious to find a treatment. Honestly speaking, in the end we really don't care if our loved one's biomarkers are up, down, or sideways. As long as their minds are clear, we are satisfied. We are saying to the field, "Look, the biology is fascinating, but I want to play Scrabble with my grandma. Don't talk to me about her amyloid burden." The first two papers recognized this reality. Then came the third paper.

THE THIRD INFLATION

Based on the first two workgroup recommendations, things were not looking too good for Dr. Gold Standard. The first two papers had basically said that pathology was only one of several things to consider in reaching a diagnosis. The third paper in the series, however, came to Dr. Gold Standard's rescue.[15] It put the pathophysiology front and center in our definition of Alzheimer's disease. More than that, however, it exploded our definition of Alzheimer's disease almost beyond recognition. This is the third inflationary event in the history of Alzheimer's disease, and unfortunately, compared to the expansions ushered in by Kraepelin and later by Katzman, this third inflation was bigger and more destructive to the field.

How could a simple 12-page paper do this much damage? It did so by "redefining the earliest stages of Alzheimer's disease." This redefinition created what was in effect a totally new stage of the disease process: preclinical Alzheimer's disease. By "preclinical" the authors meant that people with plaques in their brain (or the wrong amount of amyloid in their cerebrospinal fluid) are not healthy people. They already have Alzheimer's disease. They just haven't started to show the symptoms yet. This was a very clever jiujitsu move. The authors were attempting to turn one of the greatest weaknesses of the amyloid cascade hypothesis into a strength. In this telling of the story, the 30 percent of elderly people who have plaques but also have normal brain function are not simply healthy people with plaques. They are sick people without symptoms.

This may seem to be just semantics, but it's actually an incredibly audacious claim. About 1 in every 10 people over the age of 65 have some symptoms of Alzheimer's disease. That's 10 percent of the elderly. The other 90 percent have normal, age-appropriate brain function. But we've already learned that about a third of the people in this cognitively normal group have significant levels of plaques in their brain. Therefore, according to this new expanded definition, they have preclinical Alzheimer's disease. The authors are essentially arguing that we should increase our estimates of the total number of cases of Alzheimer's disease by threefold. And, if I might exaggerate to make a point, remember that Heiko Braak said that less than 1 percent of all brains have no plaque or tangle pathology. By this third working group's reckoning, maybe we all have Alzheimer's disease. Even the threefold inflation is an enormous change in our definition of the disease. Worse still, the authors are making this recommendation despite the fact that there are reasonable doubts as to whether or not amyloid causes Alzheimer's disease. Ah, you may say, but aren't those doubts just the rantings of a few crazed misfits at the fringes of the field? Not really. We just read about these same doubts a few pages ago in the first two papers in the series.

To be clear, the concept of a "preclinical" phase of a disease is a very useful one. We are infected by a flu virus or a coronavirus several days or even weeks before we start to show symptoms. We want to take health care measures then to protect the patient and the people that they may come in contact with. The first nodule of tumor growth is often present weeks or months before we are diagnosed with cancer. That is the stage when our surgery, radiation, and chemotherapy have the greatest chance of success. The concept of preclinical Alzheimer's disease could also be useful if but only if—the feature we are using to define Alzheimer's is unequivocally the cause of the disease. We know with 100 percent certainty that the influenza virus causes the symptoms of flu. We know with 100 percent certainty that left unchecked, a tiny nodule of dividing cells in our lung will grow until we are diagnosed with lung cancer. We do *not* know, however, with anything approaching 100 percent certainty, that the presence of amyloid will lead to Alzheimer's disease. The supremely frustrating part of this third paper is that the authors themselves clearly understand this, but they forge ahead anyways:

We acknowledge that the etiology of Alzheimer's disease remains uncertain, and some investigators have proposed that synaptic, mitochondrial, metabolic, inflammatory, neuronal, cytoskeletal, and other age-related alterations may play an even earlier, or more central, role than Aβ peptides in the pathogenesis of Alzheimer's.

I'm sorry. You cannot apologize your way out of what you have done here. Either you are 100 percent certain that amyloid causes Alzheimer's disease and are willing to defend that view to the death, or you cannot use "pathophysiology" to define a preclinical stage, or frankly any stage, of Alzheimer's disease. You cannot call amyloid buildup a disease. As we have learned, both mice and humans are very tolerant of large amounts of amyloid in their brains. They suffer few to no disabilities, even in its presence. As an instructive contrast to the argument of the third working group, consider that when cholesterol plaques build up in the vessels of the heart, we call it coronary artery disease. That's a very useful categorization. Unlike amyloid plaques in the brain, there is a significant, predictable and measurable health problem associated with plaques in the heart vessels. To be specific, people with coronary artery disease go to see a doctor because they don't feel well. Plus, we are 100 percent certain that the narrowing of the coronary arteries caused by the cholesterol deposits provokes the metabolic and finally the electrical crisis that is a heart attack. To repeat, plaques in the heart make people sick and can cause a heart attack. Plaques in the brain do not make people sick, or even uncomfortable. They are associated with risk but with no other known problem. Unfortunately, with this publication, we now had the imprimatur of the NIA and the Alzheimer's Association on a threefold inflation of our definition of Alzheimer's disease.

THE FINAL STRAW

The authors of the third paper understood that they were redefining Alzheimer's disease, and they were clearly conflicted about what they were doing. The tension in the group is clear from the language throughout the article, which is defensive and almost apologetic. In the final paragraph they admit, "The definitive studies . . . are likely to take more than a decade to fully accomplish." Said in plain language, we have this idea, but we don't have the data to back it up. Still, we are going to go with our gut, upend both basic and clinical research, and you're going to have to live with it because most of your grant money comes from the NIA and the Alzheimer's Association and their names are on this paper.

Nice lab you've got there. Be a shame if something happened to its funding.

That may be a bit hyperbolic, but I was not alone in my strong negative reaction to this series of papers. Many people accepted what the elders had handed down; they wanted to keep their research funding and keep publishing their manuscripts, but the controversy intensified. Aware of the dissension both within their ranks and in the community at large, the NIA and the Alzheimer's Association called still another working group together in 2018 and published another paper on the definition of Alzheimer's disease.[16] And again, they made matters much, much worse.

They opened their paper by stating that their goal was to "update and unify the 2011 guidelines." Without saying so explicitly, what they were trying to do was adjudicate the conflict between those who were trying to keep the clinical features of Alzheimer's disease as its defining feature and those who cared mostly about the plaques and tangles. The idea of a consensus document was not a bad one. It presented an opportunity to open the discussion in a serious way and bring real cohesion to the field. And remember, by 2018, virtually every test of the amyloid cascade hypothesis had failed. Trials of β- and γ-secretase inhibitors showed that by blocking the formation of the Aβ peptide, you made people worse. Several iterations of the antibody trials had ended or been stopped with no significant effect on cognition. The time was surely ripe for a thorough rethinking of our definitions and our approach. Sadly, two of the most important institutions promoting Alzheimer's disease research doubled down yet again on the role of amyloid. Quoting from the abstract of the paper, "Alzheimer's disease (AD) is defined by its underlying pathologic processes." Can't get much clearer than that: no plaques means no Alzheimer's. But so far this is just saying out loud what had always been the silent bias of the field. What came next was the true bombshell: "The diagnosis is not based on the clinical consequences of the disease."

The brazenness of this statement is really hard to describe. Given the paper's subtitle, "Toward a Biological Definition of Alzheimer's Disease," it was as arrogant as it was inaccurate. There is not one shred of biology in the entire 28-page document. Plaques and tangles are just descriptive pathology, not experimental biology. It was basically sticking a finger in the eye of the entire biological science community and most of the basic researchers in the field. What the authors were trying to do is say it's *only* the plaques and tangles that define Alzheimer's disease. There were soon many published critiques of this paper in the literature.[17] I will extract just one quote from Mario Garrett to describe how thoroughly ridiculous the assertions of this paper were viewed to be:[18]

A clinical disease—a disease that is experienced or has observed consequences—is now being argued to be exclusively a biological disease. But Alzheimer's disease is only important because it is a clinical disease. . . . It is of no consequence what the biology is if the disease is not experienced or observed. By reversing this truism, [and instead saying] that the biology is more important than the outcome of the disease, the authors are transforming how we look at health and disease.

That is a formal restatement of my earlier comment that the biology is fascinating, but I just want to play Scrabble with my grandma. We could go further to take apart the logic of the paper, but I think you understand what happened here. The authors handed down their unchanging verdict: without amyloid deposits there is no Alzheimer's disease. Nearly 30 years later they were basically repeating the warning of our External Advisory Board, "If you're not studying amyloid, you're not studying Alzheimer's." Garrett, in his critique, attributes financial motives to the authors, pointing out that virtually all of them have either large NIA grants, substantial PhRMA ties or both. I know most of these people personally, and while I wouldn't dismiss the possibility that their thinking was modified by their connections, I believe that they were acting in good faith as they wrote their words.

I just think they were dead wrong.

This brings us back to an important question we asked earlier: What happened to our cure? We now know that a big part of the answer is that we have been sidetracked by our obsession with amyloid and its use to define Alzheimer's disease. This is a huge problem because the definition of a disease is one of its most important attributes. Without a precise and accurate definition, there is no way to find a cure for any disease. You can't fix something if you don't know what's broken. Sadly, throughout the long history of Alzheimer's disease research, strategy and politics have overruled science in the push to apply the label of Alzheimer's disease to an ever-larger fraction of age-related cognitive decline and aging. As a result, we are left with basically with no definition—or at least none of any value. Being in this situation, we are effectively blocked from making any real progress toward treatment. For proof of this, one needs to look no further than the unbroken string of expensive clinical trial failures. Our political calculus has overruled our common sense and caused us to stop listening to our own data.

This is no way to study a human disease.

IV

WHERE SHALL WE GO FROM HERE?

===============

At this point in the book you would be justified in feeling depressed about the state of Alzheimer's disease research and the potential for a treatment ever being found. The purpose of this final part is to return a bit of optimism to our discussion. I want to illustrate some of the clear pathways forward that hold promise for real progress against the scourge of dementia. To get to a map of those pathways, however, we will first have to take a detour. We need to roll up our sleeves and work through a crash introductory course in aging. We need to do this because age is an absolute prerequisite for Alzheimer's disease, and there is no solution to the problem of dementia that doesn't rely heavily on understanding the biology of aging. It will be particularly important to see how aging affects the brain. Even if all we want to do to fight Alzheimer's disease is uncouple our brains from the aging clock that is ticking in our bodies, we still need to understand all the parts and pieces of the clock. It's the only way we can figure out how our brains "tell time." And it's why the next chapter on aging is a critical part of the book.

With a firmer background on the nature of aging we will be able to build a new disease model. Amyloid is included, as is tau, but they are now in minor supporting roles and in a much richer biological context. They will appear as important players, but not as the sole drivers of disease. With this new model in hand, we will look to our basic and clinical research portfolios and ask how a smart project manager would rebalance our resources to gain the maximum bang for our hard-earned dollars. We will look at our institutions that form important parts of the Alzheimer's disease ecosystem and see how they too need to change.

I would only caution you that the last words of the book will not be "And they lived happily ever after." There is much we can do, but in the end, there is much we will just have to accept. There is no such thing as immortality, at least on this earth. We can learn how to make our lives better and longer, but aging is not optional. The hope, however, is that by understanding the cogs and wheels of the aging process, we can look toward a day when Alzheimer's disease is a life option that we can politely but firmly decline.

10

A LAYPERSON'S GUIDE TO THE BIOLOGY OF AGING

Why and how we age is probably one of the greatest unsolved mysteries in biology. That seems counterintuitive in part because aging is so easy to recognize. We are pretty good at making fairly subtle age discriminations, especially in our fellow humans. Have a look at the six photographs in figure 10.1, and see if you can match the age with the person in the picture. These are six different people, so it should be much harder for you to match correctly than it would be if it were the same person photographed at six different ages. Also, I've used different genders and different ethnicities in my example, so your discrimination has to be much more generalized. That increases the difficulty still further. Plus, only the face is shown; there are no cues from other body features. All in all, this should be really hard.

I show this picture to my undergraduate biology classes as a "quiz" when I give my lecture on aging. I doubt you'll be surprised to learn that almost everyone gets a score of 100 percent. I'm sure you did too (and just to be sure, $A = 1$; $D = 5$; $F = 10$, $C = 20$; $B = 40$; $E = 60$ years old). What are the cues in these photographs that made the question such an easy one? There are many, beginning with the gray hair on the person in panel E. Aging happens to just about every single one of our features. Our hair thins and grays. Our skin tone diminishes because we lose subcutaneous fat, so our skin becomes less elastic and wrinkled. We develop pigment

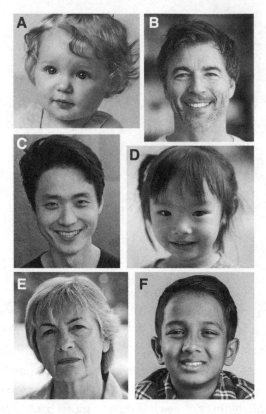

10.1 Match the age of the person with the photo of their face. There is one person represented at each of the following ages: 1, 5, 10, 20, 40, and 60.

spots on our skin. Our eyes and ears and noses lose sensitivity. Inside our bodies it's the same story. Every organ works less and less well as we get older. We lose muscle mass. Our livers and kidneys have a harder time keeping up with our body's demands. Our brains are not spared either. Our processing time slows down. Our short-term memory declines. Scientists have looked at the structure of the brain as we age, and, as you might expect, the nerve cells have smaller and sparser branches. Along with that the number of synapses (ping-squirt-ping) goes down. All this describes us as we age, but what *is* aging? Biologists have a surprising amount of difficulty in answering this question. It's so easy to see, but it's frustratingly hard to define in purely biological terms.

WHY DO WE AGE?

One of the most puzzling questions in the field of aging is also the simplest: *Why* do we age? No feature of any biological organism is maintained over the generations if it does not confer some sort of advantage to the organism. The key "currency" of this struggle for advantage is reproductive success—to be the best at passing your genes on to the next generation. As a result of this evolutionary principle, biological success is measured at the level of individual organisms, not small groups or large populations. The actions of our neighbors matter to us, but neighbors can be helpers or enemies; we can never be entirely sure. In the end, it's our own genes that we're trying to pass on, so it's those genes that count, not our neighbors'. How does aging benefit an individual? Somehow it must, because aging turns up as a feature of almost all living organisms. Any cell that has a nucleus, or any organism that is made up of many such cells, experiences aging. We age, but so do dogs and cats and flies and fish and worms. Even single-cell organisms, such as yeast, can be seen to age. Aging has been maintained not just over generations but over hundreds of millions of years of biological evolution. That means that it must confer some sort of advantage. But that seems crazy. How does aging benefit me, or anyone for that matter? And how does the answer tell us anything at all about Alzheimer's disease?

Without going into the details of evolutionary theory, the simple answer to the question of why we age seems to be essentially that there is nothing to stop it from happening. We cannot select against it. In this view, the problem lies in the fact that in the wild most organisms are dead well before they reach the end of their natural life span. For example, as recently as 500 or 600 years ago, human life expectancy was somewhere between 30 and 40 years. Probably half of us died before we were out of our teens, usually of infection. Anything that gave us an advantage after we were 50 years old wouldn't matter much, and so we would never have been able to evolve a strategy to lengthen our maximum life span, even a little bit. There would be no way to test whether some genetic modification made us better or worse at 200 years of age, because we never got there to try it out. If Alzheimer's disease doesn't start until we are in our 60s or 70s and most people are dead by their 30s or 40s, how would we be able to naturally evolve a way to avoid it? In a very practical sense, it's nothing that members of the human species ever needed to worry about. Our environments are different now, of course, due to major changes in our public health behaviors that have dramatically increased our life expectancy. The changes have happened so recently—a heartbeat in evolutionary time—that there has really been no way for our biologies to have meaningfully changed in response.

Things are probably even worse than this simple argument would make them appear. There is a theory of the evolution of aging that goes by the horrible name of *antagonistic pleiotropy*. The basic idea of this theory is that if a genetic mutation gives us an advantage when we are young and in our reproductive prime, it will be actively retained (selected for) even if it hurts us badly later in life. That early advantage will matter in the gene passing competition, but since that wonderful gene turns around and bites us later, we deteriorate faster and aging gets baked into the cake. To understand how a gene could work this way, imagine a genetic change that caused us to develop lots of extra muscle from our pool of stem cells at a young age. This would give us more strength and endurance when we are young. Our muscles would grow faster and bigger, and so we would be better at competing for food, for shelter, and for mates. Any individual with this change would be selected for because he or she would likely end up a winner in the gene passing competition. But if that genetic change did not also come with a change that increased the total number of stem cells that could be produced, we would run out of stem cells at a much younger age than our neighbors who didn't have the change. In other words, our muscles would atrophy rapidly with time. That's aging. And even though it's counterintuitive, we would actively select for this gene despite the fact that it condemned us to an early loss of muscle mass and premature muscle aging. Worse still, if a different mutation were to occur that increased the number of our muscle stem cells but did not increase their recruitment when we were young, that change would not be selected for even though it had the potential to extend our life span. In a coldhearted world, we would be dead of other causes long before the advantage of the genetic gift of extra stem cells kicked in.

So aging is baked into our biological cake. We might never have noticed except for the fact that we evolved into a clever species. We figured out how to extend our life span to a greater and greater fraction of its theoretical limit. That is very much a double-edged sword. We're living much longer, but the genes that the process of antagonistic pleiotropy condemned us to carry in our genomes get to reveal all their late-life disadvantages. Thus, Alzheimer's disease and other age-related conditions have become major problems. In this way, Alzheimer's disease actually represents a triumph of modern medicine: we have figured out how to live long enough to get it. Given time, our biology would almost undoubtedly adapt, and we could start selecting for genetic changes that extended our youthful health and vitality further and further into our adult years. But somehow kicking the aging can 100 generations down the road hardly seems an acceptable solution to current looming problem of Alzheimer's disease.

As a biologist, these insights into the aging process are important and frustrating at the same time. The important parts of the insights are the ones that bring biological logic to the many degenerative changes we see in our bodies over time. We can start to see how they are natural events that result from the passage of time. It becomes clear that they were never actually "designed" to be that way. They just happened. The implication is that aging is not one highly regulated process but rather the biological consequences of selecting for a virtual cornucopia of genes that help us when we're young. It even goes some way toward explaining why, over the years, many scholars have argued that conditions like Alzheimer's disease are not diseases at all, just exaggerations or accelerations of the normal events of brain aging. The frustrating part of the insights into the nature of aging is that if our bodies and minds are not optimized to live forever, then the logic applies equally to every part of our body. All of our systems are likely to suffer from design elements that need tweaking. The frightening implication of this idea is that we will probably need to address more or less all of these elements if we want to slow aging.

WHAT DRIVES OUR AGING PROCESS?

We can describe aging at many levels. We can describe it at the level of the whole organism, which is why the six photographs were so easy to rank by age. We can describe it and measure it at the level of the organ. We can also describe it at the level of the cell. The problem is when we're at the level of the single cell, we don't really know what aging is at all.

How does a single nerve cell know how old it is? Where does one cell's concept of time come from? Does each cell have a clock in it that ticks off the hours and days? Or is there a master clock somewhere in our body, and all our individual cells tell time by "looking" at it? We are beginning to find the answers to these questions, but we are only beginning.

We now accept that Alzheimer's is a true disease with its own biological basis, not accelerated normal aging. But given the multidimensional nature of the aging process, when a person loses mental function, we need to determine how much of the loss is due to Alzheimer's disease and how much is simply age appropriate. Fighting aging and fighting Alzheimer's disease are almost undoubtedly two different fights. But they are two fights that are so thoroughly intertwined with each other that the battle is a bit like house-to-house fighting against an army of insurgents scattered all over a densely populated city. What is needed are precision tactics that kill the processes that give us Alzheimer's disease but spare those that are parts of our normally aging brain. So, who are the biological and chemical enemies that we need to attack?

We already spoke about oxidation as a possible cause of Alzheimer's disease, so it should come as no surprise that oxidation is also a prime suspect as a cause of aging. From the perspective of the oxidation hypothesis of aging, we are all just iron bars left out in the weather. We slowly rust away. Oxidation does indeed increase with aging, but that is just proof of correlation. It says nothing about whether oxidation causes aging or aging causes (allows) oxidation. To focus on the question of causality, we need to ask why oxidative damage increases with age. It's not likely that the forces that drive oxidation increase. Going back to the aging-as-iron-bar analogy, the weather that the bar is exposed to is sometimes fine and sometimes foul. But from year to year the overall weather is pretty much the same. That suggests that what must change are our defenses against oxidation. Cells have many different enzymes and other substances that block oxidation from happening or fix it once it occurs. These repair systems are crucial to our health and to our survival. As we age, however, some of these critical components of our body become less effective. Also, sometimes a part of a cell becomes oxidized in such a way that there simply are no repair solutions. As it was with Alzheimer's disease itself, the exact nature of the relationship between oxidation and aging remains uncertain. They are correlated, but the lines of causality are not clear.

An area where the lines of causality are much clearer is the area of nutrition. In retrospect, it makes sense that a property as universal as aging would be tied into our most basic function—eating. But the nature of the effect might surprise you. You live longer if you eat less. The discovery of this linkage between food and life span started with a couple of worm biologists who set out to discover whether there were genes that could make worms live longer. The worms they were using were tiny soil worms known as nematodes. These are the scourge of many a tomato farmer, but they are a powerful biological model system. What the two scientists, David Friedman and Thomas Johnson, decided to do was to look for genes that could alter the worm life span—normally about two to three weeks.[1] To their surprise they found a gene that, when it is mutated, more than doubled the life span of the worms.

This is a bit of shock if you think about it. Normally, if you upset the delicate genetic balance of an organism by mutating a gene, you expect to see a very sick organism, or at least things should be much worse than usual. Here, you break a gene, and things get better. Doesn't this mean that the normal function of the gene is to make your life shorter? Spurred by this discovery, scientists began looking for other such genes and found them. That led to the discovery that many of the aging genes were connected to the insulin signaling pathway. At first, this made no sense. If the insulin pathway were broken, cells could not respond properly as blood glucose levels went up and down. Looking deeper, scientists found that it wasn't the entire insulin pathway that was broken by the aging mutations, only a part of it. This was a bit more comforting, and it led to the idea that what mattered was not whether the insulin pathway was on or off but rather how fast it ran. Somehow (we still don't know exactly how) if the flux through the insulin pathway is too rapid, we age prematurely,

If we can slow down the insulin pathway by mutation, what else could we do? You already know the answer, even if you're trying to avoid it. Eating a big meal loads your blood with glucose, high blood glucose stimulates the release of insulin, and all that extra blood insulin makes the insulin response pathway run faster. So, if you wanted to slow the pathway down, all you would have to do is eat less. For those of us who really love to eat, this is a frightening thought. And if you're hoping that for some reason this logic is faulty, it's not. Restricting the number of calories eaten works to slow aging, it works well, and it works in every organism that's been examined from yeast to mammals.

Don't get the idea that all we'd need to do to conquer Alzheimer's disease would be to skip dessert at dinner. The extent of caloric restriction that is needed to see an effect is quite substantial. In most organisms you need to reduce the number of calories eaten by 30 percent below the recommended daily allowance. Most of us already eat more than our recommended daily allowance, so living on 1,400–1,800 calories per day (depending on your gender and level of activity) would be more than a little difficult. Plus, we don't have a lot of data on what the secondary effects of such a large nutritional shift would be. For example, it might reduce our reserve capacity to respond to infection or environmental stress. In the end, it's not clear what impact this reduced level of intake would have for those of us who have to live in the real world.

Even with these uncertainties, this finding represented a major breakthrough in our understanding of aging, but we clearly have lots of work to do. We don't yet know if the effect is the same if you fast all the time or only some of the time. We don't know whether all calories are equal or whether it's only sugar calories or fat calories or protein calories that matter. The combinations are being dissected, and when the data come in, they will be hugely valuable. As I have repeatedly stressed, if we care about Alzheimer's disease, we care about aging—and if we care about aging, the flux through our insulin pathways should be a major clue to the biology of both.

There are other tantalizing clues that we have to consider, but for a moment let's think about the two we've discussed so far—oxidation and nutrition. Not only does each of them have a strong link to the process of aging but it turns out that they're linked to each other. The linkage is through the cellular organelles we already met back in chapter 3, the mitochondria. When our cells take up glucose, a chain of enzymes starts working on it, methodically breaking it down into smaller pieces. This generates energy, which is stored in a molecule known as ATP. Most ATP is generated in an elegant circle of enzymes that is found in the mitochondria. One run around the circular chain squeezes every bit of energy out of the glucose that we eat. To generate the ATP, the mitochondria use some pretty complex electrochemistry that really amounts to a bit of atomic legerdemain. Large multiprotein complexes start stealing electrons from the atoms in the broken-down pieces of glucose and then use the remaining protons to run a protein turbine that generates ATP. The important parts for us in all this, however, are the stolen electrons. These tiny subatomic particles need to be handled very carefully. If they get loose from their tethers too soon, they generate powerful oxidizing agents. And there's your linkage. If you lower blood glucose, you lower the amount of insulin in the blood, your cells take up less glucose, glycolysis (the breakdown of glucose) slows down, the circle of enzymes doesn't work as hard, fewer stripped electrons get loose, and so there is less oxidative damage. The pieces of the puzzle are beginning to fit together.

If you're reaching for your lab coat and telling your partner not to wait up because you're going to be working late at the lab tonight to follow up on these clues, hold on a minute. There are more clues you might want to know about. One of the most important ones is the role of DNA damage in the aging process. Almost every cell in your body has a complete copy of your genome in its nucleus. Because of the Central Dogma of Molecular Biology—DNA makes RNA makes protein—we know that if your DNA gets damaged, you're asking for trouble. If the DNA code for a gene gets messed up, the RNA will be wrong. If the RNA is wrong, the protein will be wrong. And if the protein is wrong, the cellular functions that depend on it are not going to work as smoothly as they should. The bottom line is simple: you want to keep your DNA in pristine condition. The easiest way to do that, of course, would be to put your DNA away someplace safe so that it doesn't get chipped or broken. Unfortunately, that's not an option. The whole point of DNA is to use it, and every cell in your body knows that. Every microsecond of every day our cells are using their DNA to make the RNA and proteins that they need to function. But every time you make RNA from a gene, you put the DNA of the gene at risk. You have to unpack and untwist the DNA, run the RNA-making machinery down the gene, and then retwist and repackage the DNA. We also use our DNA during cell division. Our organs are constantly replacing the dead cells of our body through a process that requires healthy cells to divide, forming two cells out of one. To do that, the entire genome needs to be unwound, copied, split equally between the two daughter cells, and then rewound. Each of these events puts the DNA in danger, and while the processes have been perfected over the millennia, none of them is perfect. Accidents happen, and DNA gets damaged.

Our cells are nothing if not resourceful, however. Not only have they made copying DNA and synthesizing RNA smooth, finely tuned processes but they have evolved an entire system of proteins that serve as DNA repair mechanics. If the DNA breaks, they fix it. These systems all seem to work fine when we're young, but as we age, the normal equilibrium between damage and repair is tilted away from repair. Rather than mere correlation, however, this failure of DNA repair and the resulting increase in permanent DNA damage is almost certainly a driver of aging—an arrow of causality. We know this from rare inherited diseases that result from mutations in the genes for the DNA repair proteins themselves. These experiments of nature have a variety of complex symptoms that vary depending on which repair protein is mutated. What scientists have realized, however, is that nearly the entire group of diseases has one symptom in common: the affected people suffer from premature aging. That means that when DNA repair is compromised in any way, an organism ages more rapidly.

So, what does all this have to do with Alzheimer's disease?

First, there is solid evidence that irreparable DNA damage accumulates in our cells as we age, and it accumulates faster in the cells of the Alzheimer's disease brain. With age and with Alzheimer's, some pieces of our DNA are lost, and some are chemically modified. We know that modifications such as DNA oxidation increase with age, especially when they occur in ways that stymie the repair mechanisms. These now permanent modifications alter the cell's regulatory balance and distort the amounts of RNA and protein that are made. Along with the modifications, more and more pieces of DNA are lost as we age—chips and cracks in our fine DNA china. Even the pieces themselves can spell trouble, as we showed in my own laboratory.[2] When a piece of DNA breaks off from the genome of a microglial cell (the cells that are part of the brain's local immune system), it can leak into the cytoplasm. When it does, the microglial cell thinks this out-of-place DNA is an invading virus. So, it immediately initiates a no-holds-barred inflammatory response. Were it truly a virus, the inflammation would be protective even though substances are released during the process that can damage or even kill neurons. Fortunately for us, the response would be shut off after the virus was defeated, and all would be well. But aging cannot be shut off, and so pieces of DNA of the microglial cell's own genome keep leaking into the cytoplasm and the microglial cells keep thinking there's a virus, so they keep the response going. This sounds just like the chronic low-level inflammation that we have already seen is the situation in the Alzheimer's disease brain. This makes our understanding of the origins of Alzheimer's disease much more detailed at the level of chemistry. But don't get complacent; there is a lot we don't understand.

Overall, what these examples make clear is that the DNA of our cells is not put away in a safe place. Instead, its continual use leads to the accumulation of damage as we age, which in turn leads to errors in its coding function. The damage changes both the quantitative and qualitative output of our genes. Not only will the RNA have coding errors (forcing the wrong amino acid to be put into a protein) but it will be made at the wrong time or in the wrong amount to meet the needs of the cell. The examples underscore, once again, the linkage among the various drivers of aging. Consider the finding that DNA oxidation is one of the types of damage that accumulate with age. This particular type of damage is common and links the accumulation of DNA damage directly to the age-related changes in oxidation and hence to nutrition. The linkage between DNA damage and the inflammatory process is another example.

This linked nature of the various drivers of aging is one of the most important lessons for you to take away from this brief primer on the biology of aging. Our bodies are designed to work well in response to an enormous diversity of external challenges. They meet these challenges effectively in large part because our organs and cellular systems are designed to all work together—how we get energy from the food we eat, how we grow new cells, and how we repair damage to our cells' working parts. The systems work well together because they evolved together. None of them evolved alone nor works in isolation. The story of the biology of aging therefore cannot be told from the vantage point of any one organ or any one cellular process. Rather than be surprised when we find nutrition linked to metabolism, inflammation linked to DNA damage, and all of them linked to aging, we should be expecting nothing less. This interconnected nature of our biological processes gives our bodies enormous flexibility in meeting the biological challenges of our lives. The problem with Alzheimer's disease, however, is that our systems were all optimized to function in our younger selves; aging was never factored into the design process. As we age, the interconnections actually work against us. Deficiencies in one system (DNA repair) interact with seemingly unrelated problems in a second (nutrition), and each exacerbates the problem in the other. This is a feed-forward situation that creates a whirlpool of brain system failures fed by many converging streams—DNA damage, oxidation, inflammation, nutrition, and other insults. This is the challenge aging poses to our brains, and it the source of the problems of Alzheimer's disease.

AGE AND THE SINGLE CELL

There is a fascinating implication that emerges from the story of aging that should be a part of any Alzheimer's disease model. The photographs of the people in figure 10.1 tell us that we can easily detect aging in an entire organism. Our functional studies tell us that aging is also easy to detect in a single organ—brain, heart, skeletal muscle, and so forth. At the same time, we just reviewed the very strong evidence that a major driver of the aging process is the accumulation of unrepaired DNA damage. Since DNA damage can only occur at the level of the single cell, however, it forces us to conclude that aging must occur on a cell-by-cell basis. The implication is that each cell must age at its own rate. For our brains that means that at any one moment, as we get older, we are dealing with some neurons that are youthful and vigorous and some that are decrepit and a real drain on the system. Our brain networks are flexible and can still function even if some of the network nodes are failing, but our overall performance inevitably declines. What we see is an average decline. What we would predict, however, is that the variability in the aging process from cell to cell and even from organ to organ should be enormous. And, in fact, that variability is exactly what we see—for example, in the graph of myelin content in chapter 5 (figure 5.4).

A strong piece of additional evidence for the idea that aging occurs at the single-cell level comes from a biological phenomenon known as senescence. In common parlance, senescence simply means old age. In biology, we use the word senescence to describe a particular cellular state that was first described in a paper by Hayflick and Moorhead in 1961.[3] The two scientists were growing human skin cells in a petri dish and realized that after a while, no matter what they did, the cells just stopped dividing. They didn't die; they just sort of sat there. They concluded that this peculiar state was "attributable to intrinsic factors which are expressed as senescence at the cellular level." Thus, nearly 60 years ago, Hayflick and Moorhead reached the same conclusion we just did: aging occurs at the level of the cell.

Over the years we have learned a lot about this curious state of cellular being. It can be caused by exhausting the capacity of a cell to continue dividing as was the case with the Hayflick and Moorhead cells. This type of senescence is called replicative senescence. But it can also be caused by inappropriate cell division such as would happen if a tumor started to form. The cell has sensors that tell it when its cell cycle machinery has gone rogue. These sensors trigger a fail-safe mechanism that actively imposes senescence on the cell. The result is the same as it was with the worn-out skin cells; cell division is stopped. This type of senescence is called oncogene-induced senescence, and it turns out to be one of our body's major defenses against cancer.

Though it probably evolved as a cancer-fighting mechanism, the link between senescence and the aging process is strong. With age the number of senescent cells increases throughout the body. Many of these were the result of replicative senescence since most of our tissues maintain a stem cell population that constantly divides to provide a source of new cells to replace old ones. Muscle stem cells are a good example. Like Hayflick and Moorhead's cells, however, our stem cell populations have a limited capacity for cell division and when they reach that limit, they senesce.[4] That means that one of the reasons our muscles start to atrophy as we age is that they have exhausted their stem cell populations. Once that happens, dying cells can't be replaced, our muscle mass declines, and the number of resident senescent cells increases.

If this were all there were to cellular senescence, the phenomenon would be regrettable but somewhat benign. Unfortunately, senescence also has a dark side that is anything but benign. As Hayflick and Moorhead discovered, senescent cells don't die. What the two scientists couldn't see with the tools available in 1961, however, is that senescent cells are not just sitting around doing nothing. They are actively taking on a totally new cell biological state that makes them truly different from their neighbors. The senescence program that kicks in tries to alert the neighboring cells to the fact that there's trouble by releasing an entire cocktail of substances into their surroundings. The cocktail release is known as the senescence-associated secretory phenotype (SASP), and it poses a real problem for the cells in the area because the cocktail includes many of the same inflammatory proteins that we learned are released from the microglial cells. The cocktail is bad for neurons and can weaken or kill them if it doesn't go away. And since senescent cells don't die, the SASP does not go away. This is another feed-forward loop that helps to drive the aging process. With the passage of time stem cell populations are exhausted and tissue regeneration slows or stops. As the stem cells stop dividing, they can enter a senescent state and adopt the SASP. In the brain this adds to the chronic inflammatory environment and weakens our brain health.

A final twist to the story underlines, once again, the interconnectedness of our body's systems. The work comes from my own lab,[5] but its implications for aging and Alzheimer's disease are significant. As we have learned, one of the well-known risk factors for Alzheimer's disease is adult-onset diabetes. The linkage between the two was always suspected to be insulin, and that turns out to be correct. With age our levels of blood insulin rise, and this starts to happen years before blood glucose becomes high enough to trigger a diagnosis of diabetes. The higher levels of blood insulin also lead to high levels of brain insulin, and the high levels of brain insulin drive our neurons to become insulin resistant. This is a huge problem because when neurons become insulin resistant, they get their signals crossed and trigger their oncogene-induced senescence program. Then, as with all good senescent cells, their SASP turns on and the cells around the senescent neurons start dying. That makes the senescent cells a lot like neuronal zombies. They are "undead," and although they don't shamble around the brain like true zombies, their SASP nonetheless reaches out to the neurons in the vicinity and starts killing them. This latest neuronal SASP cocktail adds to the inflammatory pressure that's already a problem for the neurons in the Alzheimer's disease brain and makes it worse. Think of the overlapping levels of interaction in this final story. Nutrition is involved as it increases the level of blood insulin. Inflammation is involved since the SASP adds to the chronic inflammation of Alzheimer's disease. Plus, although our data are preliminary, it looks as if DNA damage increases in neurons as they become senescent. This is an incredibly tangled web of cell-to-cell interaction that involves cells throughout the body, not just in the brain, and yet is most probably a major part of the aging story. That means it is also the story of Alzheimer's disease.

SUMMARY

We've taken a detour off the main road from the past to the future of the study of Alzheimer's disease. The resulting extension of our trip was necessary, however. Aging is an obligate part of Alzheimer's disease. Therefore, to fully appreciate the biology of Alzheimer's, we need to understand the biology of aging. We learned that aging is driven by many factors: nutrition, oxidation, and perhaps most notably the accumulation of DNA damage. The effects of aging can be seen at every level from the whole organism to a single cell within that organism. And even though the thought might be counterintuitive, it is highly likely that aging occurs at the level of the single cell and can be highly variable among them.

The overall purpose of this final part of the book is to chart a way out of the blind alley that the Alzheimer's disease field finds itself in. To map out this course correction, we need to rethink the ways in which we define Alzheimer's disease and propose new hypotheses to explain its complex biology. This latter task is an especially difficult one because of the heavy bias in the field toward models that envision a central role for the Aβ peptide. The next chapter is my own contribution toward building a new hypothesis. The foundation of my model is the irrevocable truth that without aging there is no Alzheimer's disease. This chapter on the biology of aging was a necessary first step, therefore, as it is my intention to have aging replace the biochemistry and genetics of APP and its various secretases as the biological basis of Alzheimer's disease.

11

BUILDING A NEW MODEL OF ALZHEIMER'S DISEASE

This may be the most important chapter in the book. Now that we have a foundation of understanding in the biology of aging, we can take on the task of rebuilding what the field has knocked down over the years. The repeated inflations of what it means to have Alzheimer's disease that we discussed in earlier chapters have rendered the label nearly meaningless. The dominant disease model of Alzheimer's disease, the amyloid cascade hypothesis, has not held up well to scrutiny. Clinical trials based on its premises have consistently failed and dashed our hopes that we are getting closer to a cure. The hypothesis was imaginative and compelling when it was first proposed in 1992, but in the twenty-first century it has become tired and wanting in both explanatory power and predictive value. Its dominance in the field has destroyed the very definition of Alzheimer's disease itself.

Beginning with this chapter, it's time to start building a plan for the future of Alzheimer's research and treatment. That begins with a new disease model. There is no need for us to discard the vast trove of data we already have. That would be foolish. But we will need to discard all of our preconceptions and the dogmas that go with them. This means more than simply putting aside the amyloid cascade hypothesis. A more difficult but equally important part of the rebuilding process is to develop a meaningful model of Alzheimer's disease that is both clinically and scientifically useful.

Then, and only then, can we create the working environment we need to finally make progress against "the malignancy of Alzheimer's disease."

THE DEFINITION OF ALZHEIMER'S DISEASE—THE FOUNDATION OF ALL FUTURE WORK

The three inflations of the definition have distorted our notion of what is and what is not Alzheimer's disease. We need to reestablish a common language to describe what we are working on. I propose that we return to the definition of Alzheimer's disease as solely a clinical entity. I agree with Garrett[1] that "Alzheimer's disease is only important because it is a clinical disease" and with the authors of the 2011 recommendations[2] that in reaching a diagnosis of Alzheimer's disease "the reliability of [biomarkers for diagnosis] has not been sufficiently well established."

I would begin by turning the Katzman editorial[3] around. The biological data we've assembled over the past 45 years are sufficient to cast reasonable doubts on the assertion that early-onset familial Alzheimer's disease is indistinguishable from the sporadic forms. They are different in their presentations, and we need to go back to recognizing them as two different conditions. The separation, however, need not be vast. The clinical symptoms are similar enough that we would be justified in calling them Type I and Type II Alzheimer's disease, much as is done when endocrinologists distinguish two forms of diabetes.

It's ironic that the total success of the original Katzman strategy means that we can be completely comfortable in recommending this separation. Katzman's goal, you will recall, was to expand the definition of Alzheimer's disease from defining a rare early-onset dementia with pathological deposits—the definition enshrined in Kraepelin's textbook—to a label that could be applied to all forms of senile dementia. Beginning with the CERAD criteria in the mid-1980s and continuing through the 2018 guidelines, it is apparent that this proposed inflation has been fully accepted. Senile dementia is now nearly synonymous with Alzheimer's disease, and this is true from the most advanced research centers to the most remote primary care physician's office. I propose that we accept this situation. To be honest, by now it would be far too disruptive to try to reverse course and go back to Kraepelin's original definition. But we must cut the definition free of the pathology—the microscopic picture of the brain—and rest it on the pattern of symptoms.

Using solely clinical criteria to define Alzheimer's disease brings it into register with virtually every other complex neurological disease of aging. With this in mind, the definition we set out in chapter 1 would be a workable one. A more formal definition for clinicians would be the criteria for a diagnosis of dementia laid out by the clinical workgroup in 2011[4] and summarize here:

- The condition interferes with the person's life
- It represents a decline from a previous level of function
- It is not explained by something else
- There is clear cognitive impairment
- Two or more of the following conditions are present

  - Difficulty with short-term memory
  - Loss of executive function in the face of complex tasks
  - Impairment in spatial memory and orientation
  - Impairments in language including speaking, reading and writing
  - Behavioral changes such as apathy, mood fluctuations, agitation, and so forth.

That brings us to the problem of amyloid. Historically, Kraepelin defined Alzheimer's disease as an aggressive early-onset dementia with an unusual pattern of plaques and tangles. That definition worked as long as the label *Alzheimer's disease* was to be applied only to this tiny subset of all dementias—Alzheimer's apples (red apples with worms in them). The problem with the expanded definition of Alzheimer's disease that I am proposing is there is not a 100 percent overlap between the cases of senile dementia and the cases of brains with plaques and tangles. About 30 percent of the elderly population has plaques without dementia; 15 percent has dementia without plaques. If we were completely certain that amyloid caused Alzheimer's disease, we could argue that the 30 percent group was simply a preclinical form of the disease. I cannot overemphasize, however, that we are totally lacking in that certainty. Thus, we cannot adopt amyloid or plaques as a required part of our definition.

Anticipating the objections of my Alzheimer's colleagues, let me hasten to add that I am specifically not proposing that we totally abandon amyloid as an important part of our diagnostic process. Amyloid deposits in the brain of an individual with normal cognition are associated with an elevated risk of developing Alzheimer's disease. The presence of amyloid confers about the same risk as having a single copy of the E4 variant of the *APOE* gene. That makes information about the status of amyloid deposition an important part of any diagnostic effort. But the presence of plaques and tangles should only be used to calculate risk, not define disease. After death, finding amyloid deposits during a brain autopsy should increase our confidence that the person's dementia in life was not caused by some metabolic problem such as a vitamin B12 deficiency. But not finding amyloid does not mean that a person did not have Alzheimer's disease, and finding it does not ensure that the person did have Alzheimer's disease. We can add plaques and tangles as one of the "two-or-more-of-the-following" criteria laid out by the 2011 working group. If we do so, however, I would suggest that a person meet three or more of the bullet points.

One additional problem with our new definition is that it is potentially too broad. What the science of the last 30 years has showed us is that the broad condition that was once known as senile dementia includes people with clinical symptoms that can be clearly distinguished from clinical Alzheimer's disease. These conditions include, most prominently, vascular dementia, but also age-associated dementing illnesses such as Lewy body dementia, progressive supranuclear palsy, and others. They are not only distinguished clinically; each has its own characteristic microscopic appearance. That argues strongly that there is a different biological basis to these other types of clinical dementia, and we should not be lumping all of them into a single category called Alzheimer's disease. I suggest that there is no clear answer to this problem, but I would propose that we table the discussion for now. We will be in a much better position to approach this issue once we have a working model for the biological basis of Alzheimer's disease.

A BIOLOGICAL MODEL OF ALZHEIMER'S DISEASE

We have needed a new disease model of Alzheimer's disease for many years, but efforts at building one have been effectively suppressed. The reasons are the same as those that compelled the adherents of the amyloid cascade hypothesis to exclude any thinking outside of the amyloid box. With the information we now have, and spurred on by the continuing failure of amyloid-based clinical trials, we are fully prepared to assemble a new conceptualization of the disease. What follows is one person's contribution to the struggle to move our field beyond amyloid.

Let's start our model building by first acknowledging the extraordinary complexity of the problem that we are facing. Alzheimer's disease is arguably the most complex disease to which we humans are susceptible. The brain is the primary target organ, and the impact on its structure and function is widespread: short-term memory functions, but much more. The complexity of the disease arises because the brain itself is amazingly complex. The computational level at which our three-pound brain functions is nearly unimaginable. The average human's brain has 85 billion neurons. Each of these connects to roughly 10,000 other neurons. That's 850 trillion connections—synapses (ping, squirt, ping). At any one instant in time, the activity or silence of these connections is a unit of information that our brains use to perform computational tasks of extraordinary complexity. To provide a sense of the scale of this biological accomplishment, it helps to consider one attempt to model this process in a computer. The Blue Brain is an EU-funded project that is located in Lausanne, Switzerland. It cost nearly €2 billion to create, and it models structures in our cortex known as cortical columns. Ten years after the project began, the program had a computational model of 100 columns, or about 1 million neurons. This was a phenomenal achievement because of the rich complexity of the interactions that are included in the program. To get to the scale of the human brain, however, we would need 85,000 Blue Brains. That would cost about €160 trillion to build, and when we were done, we would only have neurons, not a complete brain.

Theoretical neurobiologists have been slow to incorporate the idea that neurons are not alone in our brains. This brings into sharp focus the message of interconnectedness that was stressed so heavily in the earlier chapters and is hugely relevant to the problem of Alzheimer's disease. As Don Cleveland once remarked concerning the degenerative condition known as ALS, "It's a disease of the neighborhood." What he meant by that comment is that, at the cellular level, our brains are a menagerie of different types of cells all scattered throughout the brain but all working in unison to create what we perceive as mental activity. Alzheimer's disease affects all the cells of this "neighborhood," and this is where the disease complexity comes from.

Let's diagram a brain neighborhood (see figure 11.1). I've simplified the situation and represented only the five basic cell types in the diagram. We've talked about all of these cells to a greater or lesser extent. Note that each cell type has arrows pointing both toward and away from it. This is meant to show how the different cell types influence and support one another. Neurons send and receive signals and support from astrocytes and oligodendrocytes. Oligodendrocytes also create the myelin that wraps the neuronal axons and thus determines the speed at which messages move to the next neuron in the chain. Oligodendrocytes receive support from the astrocytes in the form of lipids and other substances. Astrocytes and microglial cells help to maintain the health and functionality

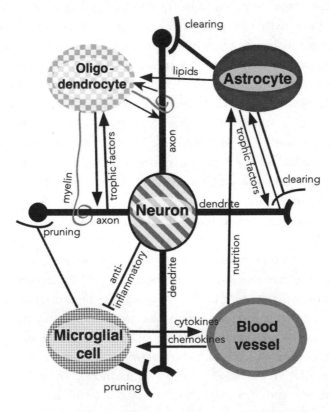

11.1 Brain "neighborhood" of five different cell types with the local interactions among the cells illustrated.

of the neuron-to-neuron synapse by clearing debris and neurotransmitters. Microglia are maintained in a noninflammatory state by neuronal influence. The cells of the brain blood vessels help transport nutrients from the blood, mainly through the astrocytes to the neurons. They also release hormones and inflammatory products from outside the brain and, in turn, take up and transmit select brain chemical signals such as the inflammatory products of the microglia to the rest of the body.

If you're thinking this is all pretty complicated, that's one important message that you were intended to take away from figure 11.1. Much of our brain's computational power can be achieved precisely because of this complexity. A second important message of the diagram is that there is much more to the brain than just its nerve cells. As rich as the network of neuronal connections might be (85,000 Blue Brains), the overall performance of the network is determined by the function of all of the cells. We've learned about the role of myelin in regulating the speed of information transfer, but that's only the beginning. Each of the interactions (arrows in the figure) has an impact on the computational network.

To help you think about the meaning of this diagram, consider the interaction between the microglial cell and the neuron. Given what you've learned, you may imagine that microglial cells do not have much to do with the moment-to-moment function of neurons. But think about what happens when you get the flu. You "feel" sick. Are you feeling the flu virus? Not at all; you're feeling the effects of substances released from your activated immune system as it attacks the virus. If we were to test you, we would find that you were apathetic and depressed, were less able to perform complex tasks, and were even having some problems with short-term memory. If you're thinking to yourself that sounds an awful lot like Alzheimer's disease, you're right. Fortunately for us all, the effects of a flu-induced cognitive decline are temporary; they pass when the virus is destroyed and the immune system deactivates. The example, however, illustrates how what we call cognition is a collaborative effort involving many different cells and types of cells. The transient dementia that is brought on by the microglial overreaction is a foreshadowing of how we need to consider all of the cells of the brain as we build our model of Alzheimer's disease.

Before we get to disease, however, let's start by building a normal brain. Figure 11.1 is a starting point. We're going to use all the cells in the diagram, not just neurons, so we'll be building a brain with colors as well as connections. Instead of just putting neurons together, let's start building brain "cities" by putting together individual brain neighborhoods. Figure 11.2 no longer has the labels on the arrows that were present in figure 11.1, but it illustrates how neighborhoods could fit together and function together. The individual neighborhoods still have a coherence, but now the neighborhoods interact with each other to widen the influence of each one. That's especially true because the diagram oversimplifies the range of the various effects. The neuronal axons can travel for great distances. The cytokines and chemokines from the microglial cells and the blood vessel cells can diffuse well beyond the confines of one neighborhood. The same is true for the nutrients that are taken up and supplied by the astrocytes. That means that the complexity of four neighborhoods is greater than four times the complexity of one.

For now, we can stop at this level of two-dimensional complexity. In a real brain, however, two-dimensional sheets of cells such as the one illustrated in figure 11.2 are interacting in three dimensions. My colleague Kai-Hei Tse helped me illustrate that concept (see figure 11.3). It is difficult to render this idea during the rest of the model building, but keep it in the back of your mind.

The next step in creating our brain model is to build a "nation" from many brain cities. That idea is diagrammed in figure 11.4. If you squint, you can still pick out the individual neighborhoods, but the overall impression is that of a textured surface with a tweed-like coloring to it. Now imagine the entire nation surface pulsating with small changes in the intensity of each region over time. The pages of a book such as the one you're holding are not able to render this next idea clearly, but multiply this image about a billion times (literally) and you would be pretty close to a conceptual model of a healthy human brain. The model is still missing all of the long-range connections that further enrich the interactive nature of the individual elements. The neurons are the strongest example of this. Relative to the scale of this drawing, they send their axons over great distances and thus impact the behavior of faraway neighborhoods and cities. To work on the model in the lab, we would need to detail just how this additional level of complexity impacted the model. For purposes

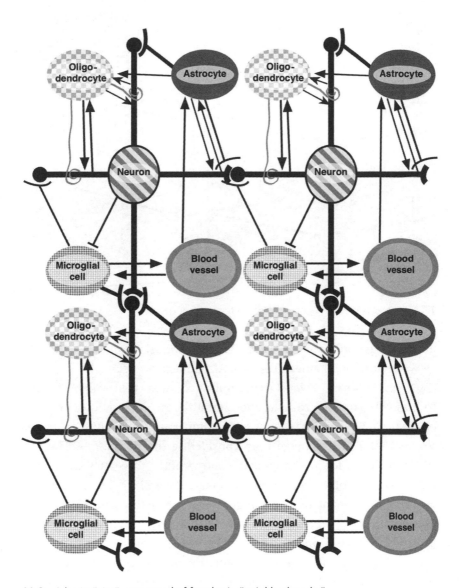

11.2 A brain "city" composed of four brain "neighborhoods."

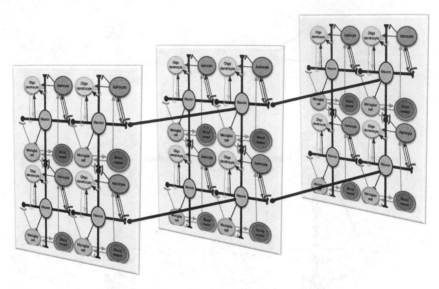

11.3   Interactions of "neighborhoods" in three dimensions.

of our discussion, however, we can keep the model simple since it carries all we need to build a new theory of Alzheimer's disease.

Even at this simpler level the complicated interactions can be difficult to grasp. An analogy to help understand the model would be to think of it as an LED display. The analogy is imperfect for many reasons, but several useful ideas emerge from considering the similarities. The LED screen creates color by changing the output of the three RGB elements— red, green, and blue—that together make up one pixel. The brain model we just built is made up of small groups of interacting cells that I called "neighborhoods." By analogy with an LED display, a brain neighborhood is the equivalent of one pixel. If an LED screen looks white, it's because the output of the three RGB elements in each pixel is equally bright. If the green elements dim, the screen becomes purple in appearance. If the red elements dim, the screen becomes aquamarine. Our brain model is more complex because instead of three elements, the brain neighbor-hoods have five different cell types in each pixel. Now let's push the anal-ogy a bit further. We don't use the LED TV in our living rooms to display a single block of color; we use it to watch moving images. The same is true for our brains. In this analogy, the coordinated flickering of our brain

11.4 A brain "nation" created by the interactions of many brain "cities."

pixels produces the equivalent of a brain movie for us. As I said, the analogy is imperfect, but it is useful.

One important idea that emerges from this way of thinking about the brain is that the "color" of a pixel is important. Ignoring the other four cell types and watching just the neurons flickering would be the equivalent of watching a black-and-white movie. Many great movies were made in black-and-white, but the advent of color improved the movie-going experience immensely. Our brains can and do use color in their computations because in each of our brain pixel elements the black-and-white output of the neurons is colored by the output of the four nonneuronal cells: the oligodendrocyte, the cells of the blood vessel, the astrocyte, and the microglia. That dimension of color dramatically enhances the computational capabilities of a normal, healthy brain.

The next step is to consider the effect of aging on the model. Remember that each cell in the array is undergoing the aging process as a separate entity. Suppose that the green astrocyte in Neighborhood No. 134,701 accumulates so much oxidative damage that it dies. That pixel loses its green element and its color changes, which would alter the computational output of any network that included that pixel. With its new color, for example, the pixel might display a fleck of red in what was supposed to be a green landscape. If the mitochondrial function of the neuron in the same pixel caused the neuron to die too, the entire pixel might go dark. Of course, a change of color or complete loss of a single pixel in a brain with billions of them is not a cause for concern and might not even be noticeable. But with time, the accumulated loss of cell functionality would cause the overall quality of the image in our brain movie to degrade. That is a pretty close description of how the aging process changes our mental capacity.

The model also helps us think about differential vulnerability. If oligodendrocytes, for example, are more sensitive than the other four elements of a single brain pixel to the changes that occur during the aging process, they will malfunction or die disproportionately to the other elements. The color of the entire brain will lose it oligodendrocyte hue (if the oligodendrocyte normally gave a pink color to the pixel, the whole brain would tend toward a more bluish tint). This is a second way in which the model allows us to capture an important feature of the aging brain.

Now we are ready to use our model to describe the biology of Alzheimer's disease. In the examples we have considered thus far, the damage to the integrity of each of the neighborhoods/pixels in our model is randomly distributed. The challenge for us is that during the course of Alzheimer's disease, the biology of the brain does not change uniformly or randomly. We lose our short-term memory because we lose brain mass in specific regions of the brain—hippocampus, entorhinal cortex, and the basal forebrain nucleus. Our behavioral changes are ascribed to losses of cells in other specific regions—the locus coeruleus and the dorsal raphe. The names aren't important. What's important is that nearly all disease models come up short when facing this challenge, including the amyloid cascade hypothesis. All of the changes that are described in the cascade are equally likely to occur in every brain region. Where does the anatomical and functional specificity come from? Why does one aging brain tip toward an Alzheimer's disease process while a second one tips toward Parkinson's disease and a third toward vascular dementia? There is no answer yet; that's part of what an invigorated program of Alzheimer's disease research needs to address. The new model, however, offers a productive way in which to think about the problem.

Let's go back to the LED analogy and change it to an old-fashioned cathode ray tube (CRT). This older display technology creates color in the same additive way as an LED, but it has one significant drawback. If the same image is left on the screen for too long without changing, a "ghost" of the image becomes permanently "burned into" the display. This burn-in might be a good way to think about how Alzheimer's disease can be illustrated in our model. Remember that each of our pixels is made up of five cell types; the burn-in could happen in any of them. The idea is that the usage of the system drives the degradation of the network properties in ways that are specific to the usage. Certain patterns of behavior or nutrition or infection could burn in a vulnerability in a disease-specific pattern. This would then interact with the overall problems caused by the aging process. The result would be an age-related tipping point where the interconnected natures of the neighborhood interactions drove a region-specific feed-forward destruction of specific functions.

Applying this to our model, imagine that one small part of our brain TV had a tendency toward a bit of epileptic activity. That would burn in a gradient of neuronal color loss to that area of the screen. It might overlap with a second area where an irritant such as a virus or a bacterium caused a small amount of extra inflammatory activity (see figure 11.5). That would burn in a striped loss of microglial color to that region. Envision this at the scale of the entire brain, and consider a case where an irritant or activity pattern was a common result of day-to-day life. The resilience of our pixels and their ability to adjust and compensate for the stress of these irritants would vary. They would be able to repair the problems when we were young, but as we aged, we will all move toward one burn-in pattern or another. This offers a plausible theoretical model for disease specificity.

Consider a second brain in which a pattern of microstrokes occurred because the person suffered from elevated blood pressure. The blood vessel cells of some neighborhoods might be exhausted by this stress, and over time, and with age, they would alter the color of their neighborhoods to give a vascular burn-in pattern (see figure 11.6). Now we are able to address the question we tabled earlier in the chapter: Should all senile dementia be called Alzheimer's disease, or are clinically distinguishable conditions such as Lewy body dementia and vascular dementia different diseases? According to our new model of aging and Alzheimer's disease, the answer is a definitive "yes and no." The model allows for two distinct patterns of brain usage to create two different burn-in patterns on our screen. For example, a failure to manage blood pressure could have a bad effect on the blood vessel cells in certain areas of the brain. The burn-in pattern would be of a different color (checkerboard in this case), and the distribution of the checkerboard regions in the vascular dementia brain would be different from the gradient and striped ones we drew in the case of Alzheimer's disease. The second diagram would be a model of vascular dementia, and it is distinctly different from the microglial/neuronal burn-in pattern that we used to model Alzheimer's disease. So "yes," vascular dementia truly is a biologically different disease from Alzheimer's disease.

There is ambiguity, however, and it is gratifying that our model actually predicts what we see in the clinic. In most cases of dementia, the microscopic examination of the brain shows a mixture of pathologies. The most common overlap with Alzheimer's disease is in fact vascular dementia. Microscopic evidence of vascular dementia occurs in over 80 percent of all cases with microscopic evidence of Alzheimer's disease.

11.5 A model of how, with age and use, a burn-in of a Alzheimer's pattern of inflammatory activity (gradient) and neuronal activity (stripes) leaves a ghost image on the "national" network.

11.6 With age and repeated overuse, a specific burn-in pattern of vascular activity becomes a permanent ghost image on the "national" network.

11.7 A brain in which the stresses of both Alzheimer's disease and vascular dementia occurred in the same brain.

That is easy to explain with our model. Aging weakens all the cells of the brain, and local injuries and long-term burn-in drive regionally specific patterns of network failure that are specific to different types of dementias but usually overlap with one another—clinically and biologically. The changes that start a feed-forward loop of Alzheimer's disease do not in any way preclude separate changes that drive a feed-forward loop of vascular dementia. What should actually be rare is for one loop to occur in the complete absence of any other. That is precisely the situation that the pathologists have seen, all the way back to Alois Alzheimer himself.

Taken together, I would suggest we now have a comprehensive model of Alzheimer's disease. The most important feature of this new concept is its focus on the combined interactions among cell types in a neuronal neighborhood as the "unit" (or "pixel") of normal brain function. The loss of normal brain function in any specific disease is not viewed as a blockage of a single linear biochemical pathway or cascade. Instead it is conceived as a distortion of the network of interactions among small neighborhoods of cells. It models the real-life complexity of the brain by proposing higher and higher levels of interactions. Neighborhoods interact in new ways to form "cities" of cells; "cities" interact in still different ways to from "nations" of cells.

One implication, if we use this model as our core hypothesis, is that questions such as which cell type kicks off the Alzheimer's disease process are probably not going to be very productive. The target of the disease need not be a cell; it can be an interaction (one of the arrows in figure 11.1). For example, imagine we wanted to test the idea that the loss of myelination is a major contributor to the symptoms of Alzheimer's disease. We could spend a lot of time asking whether the problem first appeared in the oligodendrocyte that builds the myelin, or in the neuron that must accept it, and find ourselves frustrated when our experiments to determine the answer proved ambiguous. The new model says that the question, as posed, is not the right one. The most likely situation is that the disease starts by attacking the interaction between the neuron and the oligodendrocyte. The first failure could occur in either cell and still kick off the same disease process. We would need to discover the cell biology behind the neuronal and oligodendroglial responses that blocked the interaction. But we should not try to choose between them to find the most effective new therapy; we should go after both.

In my critique of the amyloid cascade hypothesis, I argued that a good hypothesis is only useful if it is consistent with current data and offers testable predictions about how the biology of the disease leads to the observed clinical symptoms. The neighborhood model of Alzheimer's disease is no different. In this chapter, I have tried to illustrate that it is consistent with our current data set. In the next two chapters, I will lay out a proposal for using this model to adjust our thinking, our research efforts, and our institutional missions to better attack the problem of Alzheimer's disease. It is my hope that after finishing your response will be, "Now *that's* how you study a human disease."

12

REBALANCING OUR RESEARCH PORTFOLIO

We have a lot of lost time—probably 10–15 years—pursuing an amyloid-only route to a cure for Alzheimer's disease. We have a lot of catching up to do, and the task ahead of us is daunting. We must devote ourselves to what could be a substantial period during which we take the time to realign our goals with a modern biological understanding of the human brain. This will not be an easy exercise, nor will it be quick. The first step will be for the field to accept the weaknesses and failings of our current definitions and concepts of what it means to have Alzheimer's disease. The next steps will be for us to come to an agreement on a new definition of Alzheimer's and allow consideration of disease models in addition to the amyloid cascade hypothesis. The last steps will be to formulate testable hypotheses based on these new ideas, take them into the laboratory and then into the clinic, and see if they hold up. There are many shapes that this rebalancing of our research efforts could take, but the broad areas where we need to dig in are already clear.

THE BIOLOGY OF AGING

By far the most important area of our field where we need to adjust our research efforts is in studies of the biology of aging. It is in this area that our depth of knowledge is the shallowest and our need for new information is the greatest. Chapter 10 was a brief overview of the current state of the field. To advance beyond these descriptions will take a massive, well-funded effort. There are many aspects of aging that would benefit from new ideas and new talent. The lack of "curb appeal" in the study of aging that I described in earlier chapters explains why talented young researchers have shied away from the field. The work takes time, and few young people can afford to make that investment while their tenure clock is ticking. There are few role models of scientists who have prospered in this area and no professional societies to bring like-minded folks together at regular intervals. The enthusiasm of funding agencies to put money behind new studies has been minimal, and even if you did somehow get funding, the high-profile professional journals were usually not all that interested in publishing your work. This has to change, and change is starting to be seen. Below are some suggestions for where we might start with the science. In the next chapter I'll make additional suggestions that apply to our institutions.

I laid out the case that the main driver of the aging process is the accumulation of unrepaired DNA damage. The basic argument is strong, but there is work to do to fill out the picture in greater detail. To begin with, we need to determine whether all the cells in our body age at the same average rate. Oligodendrocytes might suffer DNA damage faster than other cells in the neighborhood. I used this idea in the LED analogy as an example of how the color of the entire brain nation could be changed because of the differential effects of aging on the cells of the neighborhood (see figure 12.1). No one has ever carefully looked at how fast DNA damage accumulates over the life span for any of the cells of the human brain, let alone compared the rates of different cell types. Even within one type of cell, such as the oligodendrocytes, we don't know if DNA damage accumulates steadily, every minute of every day, or if it jumps quickly from one level to the next, perhaps in response to a life event such as an illness or an accident. Also, remember that DNA damage accumulation (and hence aging) must occur at the level of the single cell. This means we need to answer the same questions on a cell-by-cell basis and not be satisfied with just average rates of decay. Finally, we need to figure out all of these issues from two angles: Why does damage accumulate, and why do the normal repair systems stop functioning? If it seems that

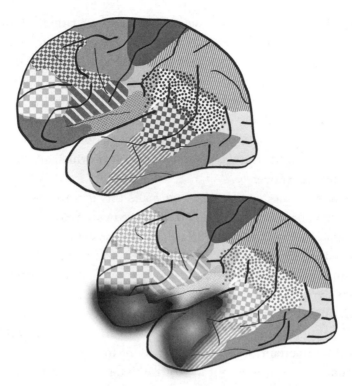

12.1 A brain "nation" that starts with a young, healthy pattern of colors but over time and under the influence of the other "nations" of the "world" develops a burn-in pattern that would lead to the symptoms of Alzheimer's disease.

we're off on a tangent here, we're not. When we roll up our sleeves and tackle questions of DNA integrity, we are not just studying aging; we are studying Alzheimer's disease.

THE AGING OF THE NEIGHBORHOOD

The new neighborhood model of Alzheimer's disease emphasizes cell-cell interaction within the neighborhoods and cities of our brains. It is equally important to remember that our brain is not an island. It is one of many organ nations that coexist in the "world" we call our body. Each of our nation systems—the pancreas, the gut, the heart, the white blood cells—ages over time, leading eventually to organ failure (or nation collapse). And just as the loss of oligodendrocytes can color the entire brain nation, a change in our hepatocytes would color the entire "nation" of the liver. And since the nations of our body are interconnected, a problem in the liver can color the entire world. In this interconnectedness, the process of aging is a bit analogous to global climate change. As our climate changes, some nations of the world may get warmer and some colder; some may get more rainfall and some less. The change is a worldwide phenomenon that is the sum of local events in the different nations of the world, and for good or for ill it creates a new environment in which all the nations of the world must operate.

To return to the human body and make this point explicitly, we've learned how too much blood insulin can drive neurons to become insulin resistant. In this way, aging in the cells in the pancreas (the major source of the body's insulin) affects the aging of cells in the brain. The implication is that, although we may try to limit our study of Alzheimer's disease to brain aging, it is equally important for us to study the accumulation of DNA damage in the cells of the pancreas as well as the brain. To a first approximation, the biology of aging is similar across the living world. That means that important groundwork can be done in experimental organisms such as mice or worms even if there are unique aspects of human aging. I admit that adding this next layer to the web of interactions makes for a much more complex network. I wish I could make it simpler, but I can't. Understanding this complexity is unavoidable if we are to understand the biology of human aging in new and productive ways.

As a basic scientist, I would love to add example after example of ways in which we should focus our portfolio of aging research projects. For the purposes of this book, however, I suggest that we skip the examples based on mitochondria, nutrition, oxidation, senescence, and more. We can do this in large measure because the big questions are mostly the same in each area. How is the dimension of time measured? Do the changes with age happens continuously or in fits and starts? How do the changes in different cell types relate to one another? These and many other questions must be on the front burner of anyone advocating for more Alzheimer's disease research. Young people do not get Alzheimer's. Understanding in biological detail why this is so will unlock many doors to dementia therapy that are currently closed.

THE BIOLOGY OF ALZHEIMER'S DISEASE AS SEEN THROUGH THE NEW MODEL

As a biologist, I view the study of aging as the single most lucrative investment that we can make in advancing our study of dementia. Over the years, however, I have interacted with hundreds of patient families and physician colleagues who are adamant that we cannot sit on our hands while basic science takes decades to build up our base of knowledge. I described this tension in the very first chapter, and it will probably never go away. The argument for action is this: even though our understanding may be imperfect, people are suffering. We cannot turn them away. We need to take our best shot on goal, and we need to do it right now. Hearing this argument, I always take the opportunity to point out that it is precisely because we have relied so heavily on this hurry-up-and-do-something approach that we are in the situation we find ourselves in today—no cure and few clues. Having said that, the families and their physicians have an important point. The urgency is real, and the cost of delay as measured in human suffering is significant. We can accept that we must do something right away, even if we are not 100 percent sure it will work.

This seat of the pants approach is probably a less efficient approach to the problem, but it is arguably a more ethical one. As I have said, certainty is a luxury that we cannot always afford. Nevertheless, I would still insist that doing something is different from doing anything. That is my cryptic way of saying that there is nothing ethical about endlessly repeating expensive trials based on old, unreliable disease models such as the amyloid cascade hypothesis. So, while we wait for our foundation of basic research to be built, let's get in the front seat of our new model of Alzheimer's disease, start the engine, back it out of the garage, and take it for a test run. Let's see if we can use it to make predictions that are testable "right now" in the clinic as well as in the laboratory

Let's start with a simple example. We've talked about the role of myelin in the function of the healthy brain and how it can begin to explain many of the symptoms of Alzheimer's disease. If we were interested in exploring this myelin-Alzheimer's relationship for new ways to combat the disease that were immediately applicable, where would the neighborhood model predict we should direct our efforts? The diagram in figure 12.2 is an

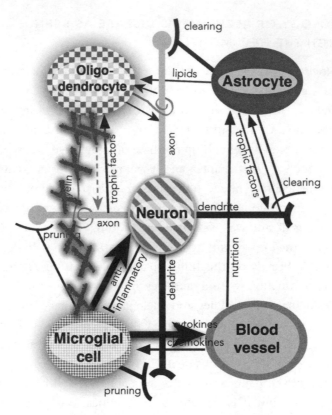

12.2 The effects of myelin breakdown (gray *x*) on the "neighborhood."

illustration of the effect on a single neighborhood of an oligodendrocyte that is functionally impaired and doomed to die. First, the biochemistry and biology of the cell itself would change. No longer able to keep up with the demands of making myelin, it would direct its resources into survival or actively orchestrating its own death. That would mean that the trophic support of the neurons in the neighborhood would stop and the myelination of neuronal axons would start to decay (illustrated by graying in the figure). This first event would be a serious problem for the nerve cell. It too would have to change its biochemistry and cell biology, directing more energy and resources into maintaining the structure of its axon. Fewer resources could be put into keeping its complex cellular architecture intact. That would likely cause connections with other neurons in other neighborhoods to be lost or pared down.

The second event, the loss of the myelin itself, would be even more disruptive. The first problem would be a loss of function. Without its myelin insulation, the speed at which the neuronal axon could send a signal to its target would go down and the energy needed support it would go up. This would increase demands on the astrocyte/blood vessel cell part of the neighborhood. Coupled with the first change, any neuronal network that involved this neighborhood would have to adjust its computational task since the information from this particular node would almost certainly be faulty—in speed if not in precision. Of course, our brain's networks are highly flexible, and the loss of one neighborhood might be nothing more than a minor inconvenience to the city in which it was located and probably not even a blip in the entire brain nation. However, were the same loss of oligodendrocyte function to occur in more and more neighborhoods, overall network function would get worse. The cognition of our brain nation would be slower, and its computational power would be reduced. That sounds a lot like dementia, doesn't it?

Worse, the loss of myelin spells trouble in a totally different direction. When a complex cell such as an oligodendrocyte dies, it doesn't just disappear. Its pieces and parts are left behind like garbage after a wild fraternity party (illustrated as the X's in figure 12.2). The brain is ready to handle the cleanup, and the cell that serves as the brain's janitor is our old friend the microglial cell. When myelin breaks down, the myelin debris is treated as garbage and the microglia eat it. But myelin debris comes with an additional feature. Besides being seen as garbage, it also triggers a very powerful immune response (illustrated as the darkening of the microglial cell and the thickening of the cytokine arrows leading away from it). This certainly adds to, or possibly even starts, the long-lived inflammatory environment of the Alzheimer's disease brain, making the inflammation problem worse. The inflammation signals also pass to the blood through the blood vessel cells. This advertises the inflammatory response of the neighborhood to the entire body. It's not certain how the astrocyte responds to all this commotion, but we can predict that it will be thrown off its game by the changes.

There are other less prominent changes that our model predicts would be the result of the loss of the normal oligodendrocyte function from the neighborhood, but let's start with just the basics and ask what the model would predict we should do clinically to fight off the bad effects of this myelin-based scenario. In other words, can we use our model "right now" to design new and improved therapies for Alzheimer's disease? The first prediction of the model is that we should do whatever we can to protect oligodendrocyte health. We could do this either by preventing their death or by stimulating their stem cells to replace the damaged ones more quickly. Sadly, because in our rush to test anti-amyloid drugs "right now," the Alzheimer's disease cupboard where we keep the pro-oligodendrocyte drugs is empty. Happily, we can check in with our colleagues who work on other human diseases of myelin loss. The most prominent of these is a condition known as multiple sclerosis (MS), and there is a lot of great biology that's been put together in that area. We can draw on all of it. Our new model clearly urges us to investigate whether any of the MS drugs that have been developed over the years might have a positive effect on the situation in Alzheimer's disease. The model predicts that some of them should. Let's find out. A dozen Phase I/II trials in this area are certainly more practical (and probably considerably less expensive) than looking for yet new ways to test the role of amyloid.

The second thing we might want to do is to block the inflammatory response that is triggered by the myelin debris. This too would be an effort that our colleagues in the MS field might be able to help us with. Current thinking about the origins of MS is that it begins as an immune system attack on brain myelin. The nature of the inflammation that our Alzheimer's model predicts would be different from the type seen in MS. The Alzheimer's disease–related microglial response is driven by the microglial cell, while in MS, the driver is believed to be a blood-borne cell type known as the T cell. The details of this difference will affect our choice of drugs, but there is no need to start from zero. That is a tremendous advantage as we face the challenge to do something "right now."

The neighborhood model also guides us toward areas where we should be investing in more basic research. These are areas that the model predicts are weak points in the system and thus vulnerable to both age and Alzheimer's disease. Learning as much as we can about these areas has clear value. Take, for example, the exact nature of the support that the neurons and the oligodendrocytes supply to one another. Learning more about the mutual biological support used by this dyad of cells would be helpful. Plus, whatever we learn will also be useful for our MS colleagues who were so kind as to let us try out their drugs. We should also learn more about the dynamics of normal myelin growth and retraction, and how the oligodendrocyte and neuron collaborate electrically to regulate normal network activity. All of these approaches are examples of work that will enrich our base of knowledge, even if they do not have immediate clinical application. That should not stop us or even slow us down. We should go into the basic science projects knowing that the risk-benefit ratio is much higher and there will be fewer solid short-term leads toward helping people with Alzheimer's disease. Nonetheless, we can start the practical approaches I suggested and at the same time continue to make important new basic science discoveries. There is no reason why we cannot work all sides of the issue simultaneously.

Before we leave this myelin example, I would offer one additional thought. Nowhere in the last few pages has the word amyloid appeared on the page. In and of itself, that's neither bad nor good. It shows, however, that we can use our new model to describe a plausible and testable path to treatment for an age-related dementia that does not require invoking the presence of amyloid or the Aβ peptide. The myelin-based pathway to dementia identifies a number of "target" areas that can be tested almost immediately in clinical trials. MS drugs that have the desired effects based on the neighborhood model and have already been shown to be safe and effective can be put to the test right away. Unfortunately, we face one important structural problem that blocks our progress along these lines. Not one of these ideas is predicted to be useful according to the amyloid cascade hypothesis. As a consequence, in the current climate, these ideas are not likely to be tested. If our experts are insisting that the amyloid cascade is the only path to Alzheimer's disease, myelin grants will face skeptical reviewers in NIH study sections, making funding for preclinical work a heavy lift. Venture capital will be withheld since the financial folks, not seeing how these ideas line up with their amyloid expectations, will balk at what would be perceived as high-risk investing. This will make it next to impossible for eager researchers to take early successes and create small start-up companies. For the same reasons there is little doubt that PhRMA would look the other way. We need to ask ourselves why this should be.

To broaden this discussion still further, I view this collision of good ideas with a stubborn backward-looking field as one of the bitter fruits of our increasingly irrational definition of Alzheimer's disease. Culminating with the 2018 guidelines, the policy of the field is now explicitly that of our old External Advisory Board: "If you're not studying amyloid, you're not studying Alzheimer's." Think of the mental knots the authors of the 2018 guidelines would have to tie themselves into if even one of the pro-myelin strategies I just outlined were to successfully treat the clinical symptoms of dementia but did not also reduce the patient's amyloid burden. It would have to be rejected as a failed trial. The people who had their cognitive decline stopped but still had amyloid present in their brains would, by the current definition, still be suffering from Alzheimer's disease. Even though I could continue to play Scrabble with grandma, if she had plaques in her brain after the myelin therapy, standard of practice would be to warn her of her expected decline and ask if she wished to enroll in anti-amyloid trials so that they could clean out those waxy brain deposits and she could finally be cured. This is the sad but inevitable consequence of a definition of disease that is based on a biomarker rather than on the clinical symptoms.

Besides myelin, there are hundreds of other ways in which the neighborhood model predicts we can rethink our approaches to Alzheimer's disease. The model is broad enough to accommodate all of the suppressed ideas that we reviewed in chapter 5 and many more. To illustrate this big-tent feature, let's use the neighborhood model to consider the role of the Aβ peptide in the advance of Alzheimer's disease. After all of the negative things I've said about the relationship of amyloid to Alzheimer's disease, it might surprise you that I would even think of including it in any forward-looking research program. That would be a misreading of my objections. What I object to, first, is the amyloid cascade hypothesis, a simple linear model of disease that does not stand up to scrutiny. My second, even stronger, objection is to the use of amyloid as a way of defining Alzheimer's disease, regardless of the clinical symptoms. While I reject both of these ideas, the data are clear that amyloid is commonly found in the brain of a person with Alzheimer's disease. We know that too much of the precursor protein from which amyloid is made (APP) can drive an early-onset form of human Alzheimer's disease. We know that too much APP in a mouse brain can lead to behavioral changes and memory problems and removing excess amyloid appears to quickly reverse the symptoms. These are important clues and they deserve full exploration.

How might APP/amyloid look if it were translated into the language of our neighborhood model? As the diagram in figure 12.3 shows, there are two currently accepted sources of amyloid (illustrated as X's): the neuron and the cells of the blood vessel. Remember that Glenner and Wong used blood vessel amyloid as their starting material and found it was identical to brain amyloid. This combined source of amyloid has two known effects on the neighborhood. The first effect is to cause transient, reversible defects in memory and behavior (the mouse data). The second is to activate the microglial cell. Though not as potent as myelin debris, amyloid stimulates the microglia to try to remove its aggregates by eating them. Amyloid also encourages microglia to mount an inflammatory response. Assuming we had never worked on amyloid before, where would the model predict we should look to find new therapeutic avenues? Getting rid of the amyloid irritant would certainly be one place we would want to look. Testing this

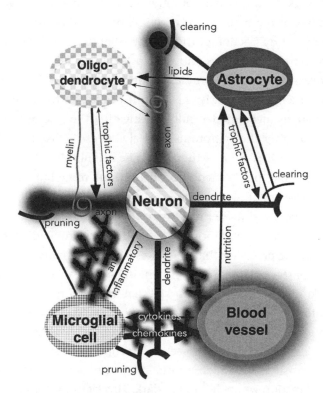

12.3 The effects of excess amyloid beta peptide production (gray X) on a typical brain "neighborhood."

as therapy was once a solid suggestion, but we've tried that approach and it has failed.

Undaunted, the new neighborhood model points to a second obvious place for us to focus: the microglial inflammation. As we learned, there have been beginning efforts in this area, but the results have been inconclusive. The drugs that have been tested are broad-spectrum anti-inflammatories—naproxen, celecoxib. Going forward, our new model would suggests that it would be worthwhile to search for agents that blocked the microglial response specifically to an amyloid signal. Also, rather than trying to separate the two conditions, the model suggests that the blood cells be brought into the model because they are also a source of amyloid. New basic science work should be done to determine the effects of amyloid on the astrocytes or the oligodendrocytes of the neighborhood. Because of our amyloid-centric and neuron-centric mindset, most work on the impact of amyloid on the brain has focused on the responses of the neuron. Experiments on the response of microglia to Aβ are but a tiny fraction of the number that have explored nerve cell responses. Fewer laboratories still have asked questions about the amyloid responses of astrocytes. Studies of the responses of oligodendrocytes to amyloid are nearly nonexistent. Even if you are a firm believer in the amyloid-is-Alzheimer's model, not investigating the role of these interactions in creating the symptoms of Alzheimer's disease represent lost opportunities to discover new therapies. The neighborhood model allows a role for amyloid, but it encourages us to think more broadly about its role in the network of interactions.

The neighborhood model makes another prediction about where we should focus in our explorations of the role of the amyloid peptide. The APP protein, the precursor that generates Aβ, is a large protein found on the surface of the cell and in its vesicles. In both locations its function is totally unknown. Many have remarked that its structure suggests it might be a receptor for . . . something. Right now, nobody knows for sure. One particular type of receptor that could fit well in the neighborhood model is one that signaled about cell-to-cell interactions. The structure of APP is certainly consistent with such a function, but because we lack any data about APP function we are left in the dark. This lack of information is one of the larger failures brought on by the insistence in the field that "if you aren't studying amyloid, you aren't studying Alzheimer's." The failure is made worse because the field has a tendency to consider the nerve cell as the only cell of any consequence in the brain. Go to any number of modern review articles, and you will see diagrams of the APP molecule sitting in the membrane of a nerve cell—often in a synapse. Even if the other cells we have discussed are all around it in the diagram, their membranes have no APP in them.[1] It might surprise you, therefore, to learn that the APP gene is strongly expressed in virtually every cell in the brain.[2] The cells of the blood vessels express just as much of it as do the neurons; the oligodendrocytes express considerably more than either of these two other cell types. Of course, to make the Aβ peptide, you would need the β-secretase, but oligodendrocytes make plenty of it. In fact, they make about three times as much as neurons do. No wonder mice treated with β-secretase inhibitors show fraying of myelin.

The broad APP expression pattern easily points the way to fresh new lines of basic science research. We could start by hypothesizing that APP, not just Aβ, was an integral part of the mechanisms of Alzheimer's disease. What if APP itself makes a cell vulnerable to damage or to premature aging when it's altered? What if the problem is not with the small Aβ peptide but with a loss of regulation of the gene for APP, causing too much protein to be made? That would probably lead to too much Aβ in the system, but are there other consequences? My own laboratory is working on the idea that APP might be a factor regulating neuronal excitability. How all of this relates to the current wealth of data we have on the effects of amyloid is uncertain.

The neighborhood model offers many other approaches based on the five cells and the interactions between them. As the myelin and amyloid examples illustrate, however, for each additional approach that we might consider, we would need to dig into the cell biological and biochemical details of the proposed system. That's important for the researchers who will lead the search for new clues and fresh discoveries. For the purposes of this book, however, let's pull back again and leave those predictions of the model for another time. Let's consider the other ways in which our new model helps us chart a fresh path toward a cure for Alzheimer's disease.

THE SEARCH FOR BIOMARKERS

Given our poor understanding of the full biological basis of Alzheimer's disease, there has been much written about the need for reliable biomarkers. We spoke about the concept of a biomarker earlier. Basically, it represents an easily measurable disease symptom that serves as a surrogate for the entire disease process. Elevated blood glucose is a biomarker for diabetes. Elevated blood cholesterol is a biomarker for coronary artery disease. Glucose is very close to the underlying disease mechanism, so it is a very reliable and useful biomarker. Elevated blood cholesterol is a little further removed, so the range of serum cholesterol values that trigger your physician's concern are a little less sharp, or at least open to discussion in polite company. Amyloid, along with tau, has been developed as almost the exclusive biomarker for Alzheimer's disease, though, as this book suggests, its use should also be open for discussion in polite company.

The neighborhood model of Alzheimer's disease offers suggestions for dealing with the biomarker question. What the new model argues is that Alzheimer's disease is a complex condition. Significantly, to get to the Alzheimer's burn-in pattern in your brain, you can travel any of several different routes. That implies that myelin debris, serum insulin levels, DNA damage, oxidative stress, and elevated levels of the products of inflammation are all worth testing. They, more than amyloid alone, reflect the totality of the conditions in the neighborhood and together should signal more definitively when the neighborhoods of the brain are tipping in an Alzheimer's disease direction. This makes the testing process more elaborate and thus more difficult to interpret, but in the context of the new model, if we cut the cord between our definition of Alzheimer's disease and the presence of amyloid, this approach makes far more sense. It certainly makes more sense than focusing on evidence for plaques and tangles alone. The more comprehensive testing regimen also means we can look for mixed dementias more easily and without apology. Elements of this broader approach have been tried, but the reliability of any biomarker, or set of markers, is usually benchmarked against the pathological picture (i.e., against amyloid), not the clinical one. We can do better.

THE AGING OF THE NATION AND ITS CONTRIBUTION
TO ALZHEIMER'S DISEASE

The model makes many predictions about how changes in an isolated pixel or neighborhood can push our brain's network function closer and closer to an Alzheimer's disease configuration. But the new model extends beyond the neighborhood and thus makes important predictions at higher levels of organization. These areas are rich unexplored sources of new ways of preventing and treating Alzheimer's disease that have been largely ignored by the field because the amyloid cascade model of Alzheimer's is completely silent about this level of complexity. The neighborhood model allows us to explore these higher levels of interactions and predict clinical interventions that may be of value.

As an example, at the level of the brain nation, several predictions emerge that have implications for ways to stop Alzheimer's disease. I presented the idea of a burn-in of a ghost image on the CRT screen as a valuable way of thinking about how a condition as distinct as Alzheimer's disease can emerge from the nonspecific, generalized process of aging. The problem that I was trying to address with this discussion is one that vexes any researcher who studies age-related dementia: Where does the regional variation in disease severity come from? Whether one thinks about the genetics of APP and presenilin or the biochemistry of the Aβ peptide, there is no real reason for any region-specific symptom to arise since the genetic mutations are identical in all cells and the biochemistry should be similar as well.

The burn-in idea tries to solve this problem by proposing that the behavior of the cities and their interactions in our brain nation is itself a driver of disease specificity. Repeatedly using the same pattern of pixels or neighborhoods over and over again causes a CRT and perhaps our brain to burn in a pattern that permanently alters the color of the output in those neighborhoods. This is a bit scary because it implies that the way we conduct our lives can alter our risk for Alzheimer's disease. It may be scary, but it's not all that much of a shock. We know that if we smoke, our risk of lung cancer goes up. We know that if we eat too many sweets, our risk of diabetes goes up. These are well-known examples in part because the pathway from the behavior to the disease is short and well understood. The compounds in cigarette smoke are powerful irritants that induce genetic mutations in cells of the lung. The wrong mutation triggers a cell to become cancerous. The sweets cause the pancreas to release more and more insulin, sustained high insulin levels make cells insulin resistant, and diabetes follows.

Alzheimer's is different, because it's so common, and because we don't know which behaviors we should increase or avoid. Because it is common, it must be true that any activities that are bad for us are pretty much standard human behaviors since one-third to one-half of us will get Alzheimer's disease if we live into our mid-80s. Because we know so little, there is not much basic biology to guide the lifestyle choices we might wish to make. Unlike the amyloid cascade hypothesis, however, the new model helps us begin to understand our options and hopefully to choose.

The model helps us map some of the positive effects we've learned about so far that come from large-scale changes in our body's chemistry. The positive benefits of aggressively lowering blood pressure in the elderly reminds us of the involvement of the blood vessel cells in the model. The increased risk of dementia caused by adult-onset diabetes is a reminder of the interconnections between our brain and our pancreas nations. If you have heard about how our behavior and perhaps dementia may be influenced by something called the microbiome (the colonies of bacteria that live in our guts), that is a reminder of the nutrition arrow that passes from the blood vessel cells through the astrocyte into the neighborhood. The epidemiology of the NSAIDs' protective effects maps onto the involvement of the microglia. The list goes on. The important message to take away from this set of predictions is that while it may seem low-tech to lower blood pressure or to prescribe a specific exercise regime, these will probably turn out to be important components of any Alzheimer's disease therapeutic program. Pairing these more systemic approaches—including nutritional ones—with more targeted approaches to the cell biology and genetics of the aging brain is very likely to be the future of Alzheimer's disease therapy. Of course, PhRMA will not be happy about these suggestions because there is no intellectual property that can be staked out around broccoli. Then again, the marketing opportunities are substantial, and PhRMA is very good at marketing.

CRITIQUING THE MODEL

As we use our model to explain the high-level interactions that lead to what I've called burn-in, it would be natural if a few alarm bells went off in your mind. I threw in the idea of broccoli specifically to ring those bells. The worry that arises is that the model is too general, too broad, and perhaps so vague that it cannot be discriminating. Without the power to tell us what makes Alzheimer's disease different from normal aging, or from vascular dementia, the model has little predictive value. This a valid criticism and concern, and it must be given full consideration.

I would argue that the model readily distinguishes normal aging from Alzheimer's disease. In the model, normal aging is conceived of as a slow, widespread change in the color of the brain nation. There may be fading, but there is no burn-in. I have been vague about the particulars of what I'm calling burn-in and exactly which colors are lost in the burned-in region. For now, it will suffice to say that any permanent distortion in the colored pattern or in its flexibility (can you turn a regional color change on and off) is the phenomenon I have called burn-in. The presence of this phenomenon of burn-in is what distinguishes disease from normal aging. But that is not enough. We need to know what distinguishes one disease from another.

Here is where the danger is the greatest that our model is too broad. In an earlier part I proposed the idea that the burn-in for vascular dementia is different than the one for Alzheimer's disease. This is very likely to be true, but it is descriptive. The model is open to the criticism that it fails to predict, in clear and uncertain terms, how the aging brain falls into each of the two different types of burn-in patterns. But is this failure a fatal weakness in the model? The question is appropriate, and the concern is valid, but I would turn it around. I would argue that the vagueness that looks like a failure of the model is actually one of its greatest strengths. That vagueness is predicted, and it accurately represents a fundamental characteristic of age-related dementias. From this perspective, the problem lies more in our definitions of disease than in the predictive power of the model. We can describe a burn-in that is a clear case of Alzheimer's disease or a clear case of vascular dementia. As we have learned, however, such clear-cut cases are rare. Most dementias are not clear-cut at all but are instead characterized by mixtures of biomarkers associated with different, supposedly separate, dementias. We already saw that the overlap between the Alzheimer's and vascular defects is as high as 80 percent. The same is true but to a lesser extent for Parkinson's, Lewy body dementia, frontotemporal dementia, and others. The neighborhood model predicts that this should be the situation. Our disease definitions, however, eschew the variation and struggle to make sense of it.

You may have wondered as you read chapter 1 why I chose the story of Dorothy as the exemplar of a person with Alzheimer's disease. I pointed out at the time that Dorothy's condition was in many ways unusual for a case of Alzheimer's disease. A geriatrician would argue that there were aspects of her presentation and history that were atypical, and even her own physician was not sure of what a final diagnosis should be. While it may have seemed odd at the time, I wanted to choose this example precisely because of the vagueness surrounding what we should call the condition that ultimately took Dorothy's life. The existence of that very vagueness is a clear and forceful prediction of the model. Aging changed the colors of Dorothy's brain in a global sense. She was a sturdy woman, however, which meant that it wasn't until she was in her late 80s that a disease process started. It began in an age-weakened brain neighborhood that burned in a pattern of irreversible changes that spread to cities and then to nations. It was a pattern that had features resembling Alzheimer's disease, but the overlap was imperfect with "pure" Alzheimer's disease. There is no question that Dorothy suffered from age-related dementia, and I propose that we call her condition Alzheimer's disease.

Our field has often used the word *spectrum* to describe the wide variation of clinical and microscopic presentations of Alzheimer's-like dementia that can be found in the clinic. In using this term, the field is making an analogy with autism spectrum disorder. The totality of clinically diagnosed autism can range in severity from relatively mild (Asperger's syndrome) to quite severe autism. As a cell biologist working on the problem of late-life dementia, having a broadly defined Alzheimer's spectrum would not be a problem. As is true for autism, we could still do excellent work on the genetics of Alzheimer's disease. For our own comfort, and to anticipate varying therapeutic strategies for different presentations of late-life dementia, we can add modifiers before or after that label as is needed. But let us agree that Dorothy's clinical symptoms were consistent with her having had something we would call Alzheimer's disease.

The Alzheimer's researchers of the world have a lot of work ahead of them. Because of the history of our field, large areas of biology that are relevant to Alzheimer's disease have remained relatively unexplored, or at least unconnected to our laboratories and clinics. That situation needs to change. The scientific tools we have at our disposal are powerful— much more so than those that were available when the amyloid cascade was first proposed. We need to pick up those tools and use them. The field is slowly moving in this direction, but every effort needs to be made to speed up the process. Because *that's* how you should study a human disease.

13

REBALANCING OUR INSTITUTIONS

Changing the direction of our research laboratories is important, but it won't be enough on its own. There are institutions that researchers rely on that also need to change. Chapters 6 and 7 described how the history of two of those institutions—the NIH and the pharmaceutical industry—helped to get us where we are today. Another institution that also needs to change is the group of professional societies that run patient support groups and under whose auspices we gather at regular intervals to exchange ideas and data. Change also needs to come to the larger ecosystem of Alzheimer's disease research. The work of our field must be communicated to the public through professional publications if it is to have any impact. Then too, what we might broadly call the media have also contributed to the groupthink mentality that keeps us from making progress. We need to look at the roles that the scientific publishing houses and the news media play in shaping public sentiment, not to mention the field's perception of itself. Change is long overdue in each of these areas. What I am offering in this chapter are suggestions for what these changes might look like. Many will seem overly ambitious. I am suggesting structural changes, which are always hard to make. Hard though it may be, we need to start an open and frank conversation and summon the courage to step out of our comfort zones. That is what will allow us to tackle the systemic issues that plague our field.

THE DEFINITION OF ALZHEIMER'S DISEASE AND THE CABAL

The groupthink that dominates the Alzheimer's disease field at present has been likened to what one might expect from a cabal.[1] It is certain that a very counterproductive crowd mentality has descended on our field. Begley and others share my own sentiment that there are no evil villains in our story; everyone is trying their best. But for a variety of nonscientific reasons— money, power, and reputation—the crowd has seen fit not only to promote its own point of view but to actively suppress the ideas of others. Even this my-way-or-the-highway attitude would be manageable if, over time, people were able to admit when they were wrong. Sadly, that is not the case.

This intransigence has had its most corrosive effects in the way it has distorted our definition of Alzheimer's disease. The 2018 guidelines[2] would not have been written except for the need to retrofit the definition of Alzheimer's disease so that it fit the new clinical trial data that contradicted the earlier definition. The only way to save the amyloid cascade hypothesis was to redefine Alzheimer's disease as a disease of amyloid. That's a circular logic that cannot be allowed to stand. Our first order of business, there-fore, must be to change the official position of the professional organiza-tions such as the Alzheimer's Association and the NIH that promote and support Alzheimer's disease research. The 2011 recommendations and the 2018 guidelines must be revamped. Much careful thought has gone into both documents, and the good parts can be retained. But the definition of Alzheimer's disease must be based on the symptoms of the patient, not on the deposits in the patient's brain. Without this return to a clinically based diagnosis, progress against the disease will never be possible. This change is not all that hard. At present, when a neurologist (or psychiatrist) says a person has Alzheimer's disease, but the pathologist says that the micro-scopic appearance of the brain doesn't have enough plaques and tangles, the pathologist wins the debate. To change our definition, all we need to do is rule in favor of the neurologist and make the clinical symptoms the gold standard of diagnosis. This is actually quite doable since the change in diagnosis would only affect the 15 percent of cases where the microscopic appearance of the brain is at odds with the symptoms found in the clinic.

We also need to rule in favor of the neurologist when the radiologist comes to us with the scan of a clinically healthy person with evidence of amyloid deposits in his or her brain. We need to abandon the idea that this person has Alzheimer's disease, preclinical or otherwise. The neurologist does the clinical diagnosis, and that is the gold standard, so the neurologist wins the debate. The person is cognitively healthy. The radiologist can add the important information that the person is at heightened risk for developing dementia, but the person can go home with his or her amyloid-spotted brain and be secure in the knowledge that he or she does not have Alzheimer's disease. This change in our definition will affect a larger group of individuals—nearly 30 percent of all healthy old persons. It will also be far more difficult to implement because it requires the field to abandon its certainty about the mechanisms behind the development of Alzheimer's disease. Difficult as it might be, however, the time has come to make the change. After all, the assertion that amyloid deposits are synonymous with preclinical Alzheimer's disease has only been with us since 2018 and has not yet entered general clinical practice. We need to make sure that it does not.

Whatever its precise language, it is important that the revised definition should be a useful one in clinical practice as well as in basic research. If we are to come together in our fight against Alzheimer's disease, we cannot afford the ambiguity of having one definition for the clinic and a different one for the laboratory. They must be the same. That means that basic researchers need an equal seat at the table where the draft of the definition is crafted.

ADJUSTING GOVERNMENT FUNDING OF ALZHEIMER'S DISEASE RESEARCH

The support that the public, through the federal government and the NIH, has provided to move Alzheimer's disease research forward has been consistently generous. Between 2012 and 2020 alone, a considerable lobbying effort, spearheaded by patient advocacy groups such as the Alzheimer's Association, has led to a more than fivefold increase in Alzheimer's-dedicated funding—to $2.8 billion from $500 million. Virtually all of that money went to the NIA in yet one more example of the success of the Butler/Khachaturian/Terry strategy to increase its funding levels. Politics of science aside, this increased commitment is impressive and a wise long-term investment on the part of the US taxpayer.

With this new infusion of funds just starting to roll through the system, the time is ripe to make a major structural change in the NIH that will pay dividends for many years to come. I suggest that we take every person, every laboratory, and every dollar of the Alzheimer's disease research effort out of the NIA and transfer them in total to the NINDS. I would argue that by almost any objective criterion that is where they belong. I tried to lay the groundwork for this suggestion in chapter 6. Alzheimer's is a disease of the brain. Notwithstanding the role of aging in our increased vulnerability and the involvement of the entire network of interactions with the other systems of our bodies, the symptoms of Alzheimer's disease are described almost entirely using the language of neuroscience. It is only logical therefore that the people who can best guide both intramural and extramural programs of Alzheimer's disease research are people who understand the normal brain, not just the Alzheimer's brain.

In this new environment, cross-pollination with other neurodegenerative diseases would be natural and much more extensive. This is particularly important in light of the new model, which predicts that the boundaries between Alzheimer's disease and other age-related neurological disorders will be fuzzy and may have less meaning than we had originally assumed. In the end, Alzheimer's disease is and should be a critically important health science priority, but it does not deserve its own institute. In this way the tail can stop wagging the dog (and hopefully, when the dog moves in, there is some high-quality flea powder sitting in one of the cupboards at NINDS so that the amyloid flea loses its grip on the tail).

That is only half of my suggestion for reorganizing the Alzheimer's disease research effort at the NIH. The second part of my suggestion is that for every person, laboratory, and dollar of the Alzheimer's disease research effort that is removed from the NIA, an equal number of new dollars needs to be pumped back in to ramp up the study of the biology of aging. I cannot state strongly enough that this second half of my suggestion is every bit as important as the first. Without it, the success of the NIA-to-NINDS transfer of Alzheimer's disease research that I just argued for will be only partly successful. These new dollars and new labs should take on problems related to cellular senescence, mitochondrial function, DNA repair and the sources of DNA damage, and oxidative damage and the body's natural antioxidant defenses. Organized under a leadership whose focus is aging and freed from vertigo caused by being wagged by the Alzheimer's tail, the revitalized agency will have a huge positive impact on studies throughout the NIH.

For those who are Alzheimer's disease partisans, this plan is a win-win. All the new information on the biology of aging can only enhance our understanding of Alzheimer's disease. For the public as a whole, this new emphasis on the problems of aging is arguably an even better investment than the money spent on Alzheimer's disease alone. As we learned in chapter 10, aging touches every system of the body as well as all of the interactions among them. A vigorous aging research program will benefit efforts across the entire NIH: cancer, heart, lung, and blood, diseases of the eye, diabetes, allergy and infectious diseases, not to mention all of the other neurological diseases of aging. Programs in each of these areas will benefit handsomely from an NIA restored to its original mission. Science and our understanding of the natural world has advanced significantly since the mid-1970s when the leaders of the early NIA needed to find some way to justify support for the study of aging. As proof of how far we've come, we need look no further than the new organization called Calico, an outgrowth of Google: "Calico is a research and development company whose mission is to harness advanced technologies to increase our understanding of the biology that controls lifespan."[3] If Google is saying we need to understand aging (life span), I don't think we need to worry any more about convincing the federal government that funding aging should be a high priority of HHS strategy.

These large structural changes at NIH need to be accompanied by a vigorous and proactive effort to encourage new, nonamyloid research into the causes of Alzheimer's disease. We desperately need fresh thinking, and, looking back, it's pretty apparent that simply encouraging alternative approaches hasn't worked very well for us. To mount a proactive effort would take a special approach to the review process. There are three NIH grant review cycles each year. In each cycle the grants received are individually assigned to one of almost 200 special review groups known as study sections. Each study section has 20–30 members and a particular scientific focus. Their targets are pretty clear from the names alone: "Biology of the Visual System" or "Clinical and Integrative Diabetes and Obesity" or "Cellular Mechanisms in Aging and Development" or "Clinical

Neuroscience and Neurodegeneration." Each grant is assigned to a primary and secondary reviewer—a study section member who is responsible for reading the proposal in detail and writing a review of its strengths and weaknesses. You may imagine that with this level of specialization, study section reviewers should be able to write a thoughtful, knowledgeable review on every grant they are assigned. They can, but with the current success rate of applications in the 10 percent to 20 percent range, the level of discrimination that is needed to rank each grant relative to all the others is quite challenging to achieve. This is particularly true when dealing with an application that doesn't fit neatly into one of the existing focus areas of the existing study sections or straddles two or three areas. The result may be that the grant is assigned to a pair of reviewers who are knowledgeable, but not about the particular topic of the grant. More insidious, if the reviewer isn't interested in a particular topic or if the grant seems to be going off in a strange new direction for the field, a grant may not generate the type of enthusiasm that is needed to reach a fundable score. This is particularly true if a reviewer believes that the question of whether amyloid causes Alzheimer's disease is settled, which most do.

We need to do something more than simply encourage more nonamyloid experimental programs. I would propose that every institute set aside a third of their review cycles and specifically exclude amyloid or tau-based applications. For example, I would propose one entire round of applications to NIA and NINDS be designated as tau- and amyloid-free zones. APOE applications would be allowed, but not if the project focuses on amyloid transport or clearance. A project on APP biology is OK, but not its proteolytic processing. The problem we face now is that as resources become limited, study sections become more conservative. Ranking two grants relative to each other is often a case of comparing apples and oranges. And since the competition for NIH dollars is fierce, just a little less enthusiasm on the part of either the primary or secondary reviewer can move the grant down below the pay line. If an innovative but risky nonamyloid grant goes to the same study section as a safe grant from an established amyloid or tau lab, the study section will typically play it safe. The end result has been that our field continues to be a monoculture. If all amyloid and tau applications were forbidden in a fraction of the study sections, we would finally see true diversity in our project portfolio grow.

The only way that we are going to achieve the diversity of ideas that we so desperately need is if certain pots of money are set aside where amyloid/tau applications do not compete head-to-head in the same review session. Who knows? Maybe even traditional amyloid labs would start to pursue alternatives that they currently do not.

ADJUSTING THE ROLE OF THE PHARMACEUTICAL INDUSTRY

It cannot be stated too often that our industry partners are a critical part of our mission to combat Alzheimer's disease. I've said before and I'll say again, all the basic science in the world is worthless if someone doesn't reduce it to practice. Yet for the past several decades the pharmaceutical industry has benefited handsomely from the increased federal investment in Alzheimer's disease research but has neglected its responsibility to help with the support of basic research. This is not to ignore the contributions the companies make by supporting the lion's share of the complex and expensive clinical trials required to bring a new drug to market. They also mass produce and distribute the new products to the hospitals and pharmacies of the world. Acknowledging these contributions, PhRMA's profits are high from their successful marketing strategies and pricing flexibility, and they save tens of billions of R&D dollars by piggybacking on the NIH-funded university laboratories that do a good part of the foundational basic research that underlies their most successful drugs.

Convincing the industry to do more to support our basic research efforts would be a straightforward case if the drug companies were true partners in the fight against Alzheimer's disease. They are not. They are driven by market forces and responsive to profit margins. Let us acknowledge that they are valuable contributors to the fight against Alzheimer's disease but would turn tail and run if they had to do it all on their own. We must be clear-eyed about the fact that the folks in the front office really don't care if we ever find a cure for Alzheimer's disease as long as they make money. That may be harsh, but it's not wrong. To be completely fair, I have known many researchers in the PhRMA laboratories of the world who are dedicated and care deeply about the valuable work that they're doing. But these dedicated people are not the ones making the big strategic choices, the go/no-go decisions about whether to close an entire research effort or to take a drug candidate to the next level of clinical trial. I argued in chapter 7 that these firms had all the data that they needed to make the correct decision and pull the plug on the anti-amyloid approach to Alzheimer's disease therapy. It would have been a sound business move, but short-term thinking drove them to give more weight to the value of their stock than the quality of their science. As a result, they have contributed to the situation the field is in. Our goal must be to work with PhRMA but keep their contributions in perspective and measured by our own more holistic metrics. To do that, we must use money as a lever as it is unlikely that any other incentive will work.

I propose that we reexamine the basis of the partnership with industry. Both academia and government need to take a hard look at their relationship and pull back. Over time, first academia and then government has become more and more involved in advanced clinical trial funding. I can speak to the academic side most directly. Using their engineering schools as models and driven by the huge windfalls that they imagine will come from developing a blockbuster drug, many of our universities have reordered their priorities based on dollars rather than on their traditional missions of education and research. I have watched over the years as talk of "knowledge transfer" or "translational research" is used to justify investing resources in large offices filled with tech transfer gurus and intellectual property lawyers. In the life sciences, at least, I doubt there are more than a handful of universities where that diversion of resources has paid off, or even broken even.

The NIH has also been bitten by the blockbuster bug. The motives are different; in this case they are anxious to prove that the large public investment that's been made in their programs over the years can consistently deliver real public health benefits. We are constantly reminded that it's the National Institutes of *Health*, not the National Institutes of *Basic Research*. Yet here too, the imagined glory that would come from playing a key supporting role in the development of a cure for Alzheimer's disease has caused the planners in Bethesda to commit tens of millions of dollars to the support of advanced clinical trials. This must stop. The government-to-industry transfer of funds that comes from NIH helping PhRMA bankroll ill-advised human trials, when added to the billions of dollars that the companies save each year by not having to support the full cost of their R&D efforts, represents something pretty close to corporate welfare. We must not let our eagerness to find a cure for Alzheimer's or any other disease drive us to spend precious basic research dollars in this way.

That puts us between a rock and a hard place. As we discussed in chapter 7, if the pharmaceutical industry did not exist, we would have to create it in order to bring new drugs to market. We need them, but we also need to be mindful of the fact that their core mission is making money, not saving lives. How shall we adjust the partnerships to cushion both the rock and the hard place as much as possible? Let's consider the steps of the drug development process. The partnerships can and do work well in the preclinical space, meaning disease-related research that has no specific application. Historically, the industry has been pretty stingy about helping out financially with this part of the process. In fairness to them, why should they? With Alzheimer's lobbyists successful securing a quintupling of the NIH Alzheimer's disease research budget, they have no incentive to add dollars to the basic research pile. But in this area of exploration, before there is any intellectual property to fight over, the process of discovery benefits everyone. That means that everyone has a responsibility to support the work financially. We can compromise on the details—academia can commit to doing work that's a bit more focused on pharmacologically useful projects; industry, with its profit margins north of 14 percent, can afford to form a consortium and invest a bit more serious money in basic research. It might fall on deaf ears in the accounting department, but there is a moral obligation to do more here.

The partnerships get trickier once intellectual property is secured. When that happens, academia and government are partnering with a single company, not a consortium. Nonetheless, the past years have shown that these partnerships can work well in Phase I trials of safety and early Phase II trials of efficacy. Sharing in the costs at these stages still makes sense from the public's point of view. As outlined in chapter 7, much of the work is discovery based, answering the question of whether the new drug works safely in humans. Industry, academia, and government are all stakeholders in these first efforts at drug development. This is also an area where the public has a stake in pushing for the test of a new treatment of a disease they care about.

Once the Phase II trials are over, however, government and academia need to hand the entire drug development process off to industry. Stated simply, the entire cost of every Phase III trial should be borne by industry alone, full stop. I make this recommendation not to save money but rather to use money to try to bring a bit of sanity to the drug development process. The fear of falling stock price will still drive a few foolish decisions to proceed where caution should prevail. If the company is flying solo, however, the financial risk will be borne solely by the folks who are making the call, and not by the taxpayer. Nothing engenders caution more than the prospect of big losses over an ill-advised call to take weak Phase II results on to Phase III.

The logic is pretty clear. If a new drug target performs well in its early trials, proving itself safe and effective in humans, no additional incentive should be needed to keep the development process going. Company executives, outside investors, trial site coordinators, literally everyone would be lining up to push the Phase III trials to get started. On the other hand, if the effects are small or present only in a subset of people in the trial and there is debate as to whether the results are statistically significant, we should take a step back. We should ask ourselves whether we really want to spend 10 times as much money on the next phase of testing or go back and recheck our preclinical work. This is actually a tough decision. My personal preference as a basic scientist is to go back and reassess, but I acknowledge that a case can be made for forging ahead. From the public's point of view, our nearly complete lack of treatments for Alzheimer's disease is frustrating in the extreme. They have been very patient, waiting for the scientists and drug companies to make progress. The frustration is high enough that having even a small-effect drug, if it could be made available soon, would be better than waiting years for a revamped trial design to make its way through the regulatory landscape. That's worth doing once, maybe twice. It's not worth doing 30 times. We should all be smarter than that.

The financial and legal responsibilities of the industry preclude its member companies from ever becoming true partners in the fight against Alzheimer's disease. Given that the true motivation of the industry is profit, not human health, we can forge a more equitable partnership by approaching it in cold financial terms—terms the industry will understand. Universities could start the process by tracking their faculty member's consulting activity more closely. Most faculty contracts specify that up to 20 percent of a professor's time can be spent on outside professionally related activities. The number seems about right to me (being a professor myself, I acknowledge a potential bias in my opinion here), but I doubt that many institutions track their faculty members all that closely. My goal in suggesting enhanced tracking of outside activities is not to burden or punish the faculty for offering their advice to industry. My goal is to implement a more audacious change. All universities should insist that any company using the expertise of one of its faculty members for commercial benefit would need to pay indirect costs on all remuneration—expenses, honoraria, everything—given to that faculty member. This would apply to all the consultants and advisory board members. Charging indirect costs is done for granting agencies all the time. The payment serves as a recognition that there are institutional costs incurred in creating a research environment that provides and maintains facilities and meets any and all regulatory compliances that would not be needed if no research were done. For example, when NIH awards an investigator $100,000 to conduct a research project, it typically pays the university an additional $50,000 or more in indirect costs. It is already the case that when drug companies collaborate with faculty on directed or contract-based research in university laboratories, they pay the university some level of indirect costs towards the general support of the laboratory infrastructure. The analogy is direct and compelling. In the case of a faculty member's consulting activity, the university is releasing up to one-fifth of the cost of a faculty member's salary plus the lost time for committee and other administrative service. This lost effort would not be needed if the company were not using that faculty member's time. An indirect surcharge of 50 percent seems perfectly reasonable. It acknowledges the contribution of the university, not just faculty member, to the consulting arrangement, and it makes the industry more of a true partner in the relationship.

ADJUSTING THE ROLE OF THE ADVISORS

This brings us to another area in which the field needs to do a bit of soul-searching. In chapter 7, we looked at the question of why the industry kept forging ahead on amyloid-based trials long after their futility should have been recognized. I suggested three answers, stubbornness, greed, and bad advice. The bad advice answer needs a different type of solution. The companies' scientific advisors—consultants and board members—were cheering the blind pursuit of amyloid-based mechanisms of Alzheimer's disease as loudly as anyone. They might have been given a pass in the early years, but that coupon expired years ago. The question is how to guard against this type of bad advice in the future. Companies choose the people from whom they want to seek advice from our most prestigious academic institutions. They are usually, but not always, physicians with deep experience in the Alzheimer's disease field. For the most part, they also have had prior experience with the drug industry, so they are familiar with its needs and its culture. They are also paid well for their advice. This has all the ingredients needed for a corrupt patronage system, but I would argue that truly serious abuses are rare and correcting them (while important) would not fix the problem. These are precisely the people who should be advising the drug companies. PhRMA is wise to choose them; they and we should be grateful that they are willing to serve. But there are adjustments that might help guard against the bad advice problem we have identified.

PhRMA chooses scientific advisors based on such sensible criteria as the prominence of the individuals, the prestige of their university, the extent to which advisors have had prior experience with the industry, and the depth of their knowledge with respect to the biology and medicine of the disease—Alzheimer's disease in our case. I would suggest that without dismissing any such sitting members, new ones be added. There are people out there who are deeply knowledgeable about the brain and the aging process but fail to meet the rest of the criteria I laid out above. These people are both young and old, physicians and basic scientists, familiar with the drug industry and not. These people should be brought onto all of the advisory councils. It has been my experience that most smart people who come to the Alzheimer's disease field from outside the field react in exactly the same way when I introduce them to the thinking in the field. Nearly all of them say to me, "But that doesn't make sense." PhRMA badly needs this attitude in their boardrooms when decisions are made about moving a drug candidate forward. Not just one lonely voice, but a diverse and full-throated chorus.

This change may sound simple, but my experience is that this will be the hardest to implement. Paying indirect costs on consulting fees would amount to small change in the budgets of most of the medium to large firms—one more cost of doing business and not a big problem. Listening to dissenting voices with an open mind is a major cultural shift. I have one small firsthand experience with this problem based on an invitation I received to advise drug companies on their science. I was invited to a pre-rollout meeting of a new drug. The company had prepared a large packet with the underlying science and pages of marketing plans. I had the foolish idea that they actually wanted us to read what they'd given us. I looked at the claims in the marketing material and then looked at the science. At one point in the discussion I raised my hand and offered the opinion that the science behind one of their claims was pretty weak and I thought it might be a good idea if they backed off on this assertion. There was an awkward silence before we moved on. Needless to say, I was never invited back to consult for that company. What I offered was constructive, honest criticism, but comments such as mine were clearly not part of the script.

ADJUSTING THE ROLES OF OUR PROFESSIONAL SOCIETIES

The professional societies centered on Alzheimer's disease and other late-life neurodegenerative diseases can change as well. Most were originally founded as patient support and advocacy groups, but many, including the larger ones in the United States and the United Kingdom, have recently devoted much of their energies and fundraising toward support of research. They also run international meetings where researchers from all over the world can get together to hear presentations on the latest developments and present their own ideas. The Alzheimer's Association even publishes a very prestigious professional journal, *Alzheimer's & Dementia*. The scientific programing of events that are organized by these societies suffers from much of the same groupthink mentality as I described earlier. This is not surprising in the least since the leadership of these groups will be the same experts that PhRMA and other groups have relied on for advice over the years. The programs of the meetings and the contents of their journals need to change to reflect the full range of views in the field. The meetings cannot be all about science; their contributors and supporters will rightfully insist that the meeting continues to have a strong patient-centered focus. That said, there needs to be more platform time devoted to educating the members of these groups about the rapidly developing basic science landscape. Contextualized in medicine, new developments in the science of aging, immunobiology, DNA damage, myelin biology, mitochondrial dynamics, and many other topics need to be presented. The information cannot be segregated. It is the instinct of most program committees to take all the basic science and lump it into separate symposia or short talk sessions that the clinicians can then avoid. I would urge instead that world expert talks, in fields related to but separate from Alzheimer's disease, be integrated into as many must-see sessions as possible. The mixing is important. Clinicians need to hear a thoughtful discussion of the news from the basic biology labs every bit as much as our molecular biologists need to hear a detailed description of the criteria for clinical diagnosis. It's one disease, so even as our laboratories and clinics are increasingly specialized, we need to remain generalists as much as possible and learn about every aspect of Alzheimer's disease.

ADJUSTING THE ROLE OF THE PROFESSIONAL AND LAY MEDIA

A final series of adjustments can and should be made to the professional journals that communicate our science to other scientists and to the lay media outlets that communicate our science to the general public. Adjusting the professional journals should be the easiest. Reviewers and editors of both high-impact, high-prestige journals like *Cell*, *Nature*, or *Science* and broad open-access journals such as *PLOS ONE* need to start holding our field to a higher standard. Inclusive open-access journals set as their goal the publication of quality workmanlike science regardless of its "newsworthiness." They should stay the course and change the least. But flagship publications such as the CNS trio (*Cell*, *Nature*, and *Science*), the *Journal of the American Medical Association*, *Lancet*, and others need to change. In these publications, an article must be both flawless in its science and captivating in its topic. Their editors must sit up straight in their chairs and take their red pens to all of the Alzheimer's disease articles that pass their review process. If a submitted paper on any other topic lacked a detailed cell biological or biochemical mechanism behind the effect being reported, these prestigious journals would send the manuscript back to the authors with polite suggestions to send it "a more specialized journal." The same needs to happen with amyloid or tau papers. It is totally unclear to me why the folks in the Alzheimer's disease field get a pass on this critical feature of their articles. I once asked a symposium audience of over 200 people if anyone in the room could tell me exactly how amyloid kills a neuron. Silence. "How about tau?" Silence. I've also asked people one-on-one just in case they were shy about speaking up in public. I still get nothing but speculations or a theory. What this means is that for all that we are assured that amyloid causes Alzheimer's disease, we really don't have the foggiest idea of how that might work at a cell and molecular level. That might have been OK for publishing in the 1990s. It's not OK now. If amyloid or tau is used as a stimulant in an experiment, the editors should insist that the authors describe the molecular mechanism by which these stimulants exert their effects. That's the standard used for papers on most other topics. Our field needs to be held to the same standard.

The lay public also deserves a more critical press corps covering the field for them. As you no doubt understand by now, Alzheimer's disease is a complicated topic—scientifically, clinically, and politically. The press, by and large, does a great job of explaining the complexity. But beyond their explanations, the press needs to start asking us the hard questions we should have been fielding for many years. As of now, most reports are based on the comments of a few experts at one prestigious university or another. That works well in most circumstances, but the probability is high that the person a new reporter speaks with is a member of the cabal; at the very least, they have their own agenda. I fully understand that most science reporters cover many different topics and it's hard to learn the detailed complexities of each one. But the Alzheimer's disease field is overdue for a grilling. Outfits such as STAT+ are starting to do that work. For the other outlets, consider this: You would never do reporting from Afghanistan without reading up on the country's cultural history. Treat the biomedicine of Alzheimer's disease as a war zone when you go in. There are alternative points of view out there; go find them.

SYNOPSIS

This list of suggested changes might be expanded, but these are the broad
areas where change needs to happen. Changing the direction of scientific
and medical research is important. The NIH and large private foundations
such as the Alzheimer's Association have an outsized leadership role to
play here. Private industry has been effective, but it needs to take more
responsibility for its own mistakes and give more than token acknowl-
edgement to the ways in which it benefits financially from the work
of basic researchers and their NIH-funded laboratories. But changing
research policy alone will not be enough. Public attitudes need to change
as well. This can only be done if all of the institutions in the fight against
Alzheimer's disease are willing to do their part. Rather than just marketing, this change in what I'm calling public attitude includes the ways in
which the field views itself: the myths, the legends, and the "it-is-well-
known-thats" need to adjust to fit the data of the twenty-first century.
That will truly prepare us to properly study this human disease.

14

FINAL THOUGHTS

The history of Alzheimer's disease research is a complicated one. In this it is probably no different than any other major human undertaking. There are elements of the story that are uplifting. There are elements of the story that are legitimate sources of great pride for the scientists, clinicians, and others who have devoted their careers to untangling the twisted strands of this most perplexing human malady. And there are other elements of the history of Alzheimer's that are not so pleasant. However, just because they are unpleasant does not mean they should be erased from the record.

The people who work in my field are smart, innovative, and passionate in their desire to rid the world of Alzheimer's disease. If you look back over the history of Alzheimer's research, I hope you feel as I do that most of the science I described was high-quality exciting research. The four discoveries we discussed in chapter 1—the sequencing of the amyloid peptide, the discoveries of the Alzheimer's disease genes, the creation of the mouse models, and the discovery of a vaccine for amyloid—were tremendous accomplishments. They stand as tributes to the teams who made them. More recently, the extraordinary technical and scientific advances in our ability to image the human Alzheimer's disease brain have revolutionized the types of questions we can ask. This and a thousand other new tools mean that the pace of discovery is not likely to slow anytime soon. So, stand up, fellow researchers, and take a well-deserved bow. I may disagree with some of your interpretations, but, as I have stressed throughout, the data we have are solid. We should celebrate them.

That said, we cannot look away from our unpleasant mistakes. They are as much a part of our story as the successes. It's whether we can admit those mistakes, whether we can learn from our failures, and whether we can adjust our course and restart that determine whether or not we can be truly great scientists. The history of Alzheimer's research is also a story of how, in our rush to find a cure, we have gone some distance down a blind alley where we have lost our way. For too long, decades if we're honest, we have focused more on salesmanship than scholarship. The need for rebalancing the amyloid cascade hypothesis and incorporating the wealth of other worthy ideas about the nature of Alzheimer's disease has been clear for at least 40 years. Yet the suppression of the other ideas has continued. The mystery is why. How did the momentum behind a single theory, the amyloid cascade hypothesis, become a steamroller intent on crushing any and all alternative models that were being debated at the time? The mystery is all the more confounding because the alternative views of Alzheimer's disease are compatible with most, if not all, of the ideas behind the amyloid cascade hypothesis. There were countless opportunities to form comprehensive models that joined ideas together and created richer, more broad-based models—opportunities that were missed. I have tried to present this puzzle from all angles, and while some pieces are fitting together, I do not feel yet that there is a fully satisfying answer to this question.

We all want to see a cure for Alzheimer's disease become available as quickly as possible. All of us as citizens have a huge stake in this. Our tax dollars pay for most of the basic Alzheimer's disease research. Our health care dollars pay for the treatments we have now and will pay the costs of any future treatments. What goes on in the laboratories and clinics matters to all of us. There is controversy in the field, but in and of itself that is nothing to worry about. Indeed, my disagreements with my colleagues are numerous and often spirited. Yet I remain confident that we are all on the same team. In Alzheimer's disease we share a common enemy. Each of us is completely certain of the validity of our own views as to how best to defeat the enemy. And though our views may differ dramatically, in my opinion this is healthy. It represents science at its best. It is in the fires of such debates that we will forge the strong steel needed for the swords we will use to finally defeat Alzheimer's disease.

It may be a bit messy, but this is how you study a human disease.

As exciting and hopeful as these thoughts may be, however, this is also a troubling time. In many ways we are entering new territory here. As we do, the Alzheimer's landscape looks different from where we started with the story of Dorothy. The light seems somehow odd, and there are strange new sounds all around us that seem vaguely threatening. Change can be scary, but stay strong. We have to make this journey. We have to find out whether the new territory we are entering truly has the rich soil we need to plant our data and reap a new crop of treatments and therapies. Come with us. Bring your photos and memories of your loved ones with you, the Dorothys of your life. We must never forget them or the people who cared for them. We cannot bring them back; we cannot even reverse the changes in those who are still with us. But we can honor their lives and their memories by being informed and critical. We can demand that our scientists and clinicians and institutional partners check their egos at the door and get about the business of finding the treatments that have eluded us for so long.

In the last few chapters, I have laid out new models of aging and of Alzheimer's disease. I have made dozens of suggestions for scientific and institutional change. Let these serve as a first crude map of our new world, a world where we can break the long string of failures we've suffered. Like many first maps, this one will no doubt be refined and polished over time as we explore our new surroundings more fully. I am optimistic that we can do this, but my optimism is not boundless. Aging and the changes that go with it will be as much a part of human existence in our new land as they were in our old one. We cannot stop it; I'm not even sure we can slow it down; but I am confident that we can learn to increase the quality of our allotted life spans, extending good health far into our old age. This is how we can make this new territory a place where we can ease the suffering that goes along with Alzheimer's disease and make sure that speaking those two words will no longer send shivers down our backs.

ACKNOWLEDGMENTS

The list of people whom I need to thank for the discussions, arguments, and brainstorming sessions that inspired me to write this book would be a book all by itself. Every interaction planted the seed of a new idea or sharpened my arguments in defense of an old one. Indeed, I value the bitter fights almost as much as the long discussions over beer and wings. Special thanks to the people who were kind enough to spare time to let me interview them for their stories and perspectives—both laypersons and professionals. We agreed that we would speak under the Chatham House Rule. So, I will not identify you as sources. That way you will never be burned by association with me and my ideas. I will offer a tip of the hat to my cousin Janet Samuels, who helped me proofread the galleys, and to my research assistant, Candice Kent, who was a big help with some of the figures. A special shout-out goes to the MIT Press and Executive Editor Robert V. Prior for taking a risk with me and publishing this book. Then there are my family and friends who sat through my rants, helped me make my points more accessible to nonscientists, and generally put up with me during the long gestation period of this book. I owe you bigly. Finally, I want to acknowledge my friend and long-time lab mate, Gary Landreth. He deserves most of the blame for getting me into this field and a lot of the credit for keeping me sane and centered while I struggle to make sense of it all.

NOTES

CHAPTER 1

1. Alzheimer's Association, "Alzheimer's Disease Facts and Figures," *Alzheimer's & Dementia* 15, no. 3 (2019): 321–387.

2. M. Ycgambaram et al., "Role of Environmental Contaminants in the Etiology of Alzheimer's Disease: A Review," *Current Alzheimer Research* 12 (2015): 116–146.

CHAPTER 2

1. This and other quotes are as translated in K. Maurer, S. Volk, and H. Gerbaldo, "Auguste D and Alzheimer's Disease," *Lancet* 349 (1997): 1546–1549.

2. H. Hippius and G. Neundorfer, "The Discovery of Alzheimer's Disease," *Dialogues in Clinical Neuroscience* 5 (2003): 101–108.

CHAPTER 3

1. G. D. Cherayil, "Fatty Acid Composition of Brain Glycolipids in Alzheimer's Disease, Senile Dementia, and Cerebrocortical Atrophy," *Journal of Lipid Research* 9 (1968): 207–214.

2. S. C. Vlad et al., "Protective Effects of NSAIDs on the Development of Alzheimer Disease," *Neurology* 70 (2008): 1672–1677.

3. E. Boelen et al., "Detection of Amyloid Beta Aggregates in the Brain of BALB/c Mice after Chlamydia Pneumoniae Infection," *Acta Neuropathologica* 114 (2007): 255–261.

4. D. K. Kumar et al., "Amyloid-β Peptide Protects against Microbial Infection in Mouse and Worm Models of Alzheimer's Disease," *Science Translational Medicine* 8 (2016): 340ra372.

5. M. Yegambaram et al., "Role of Environmental Contaminants in the Etiology of Alzheimer's Disease: A Review," *Current Alzheimer Research* 12 (2015): 116–146.

6. D. A. Drachman and J. Leavitt, "Human Memory and the Cholinergic System: A Relationship to Aging?," *Archives of Neurology* 30 (1974): 113–121.

CHAPTER 4

1. G. G. Glenner and C. W. Wong, "Alzheimer's Disease: Initial Report of the Purification and Characterization of a Novel Cerebrovascular Amyloid Protein," *Biochemical Biophysical Research Communications* 120 (1984): 885–890.

2. G. G. Glenner and C. W. Wong, "Alzheimer's Disease and Down's Syndrome: Sharing of a Unique Cerebrovascular Amyloid Fibril Protein," *Biochemical Biophysical Research Communications* 122 (1984): 1131–1135.

3. J. Kang et al., "The Precursor of Alzheimer's Disease Amyloid A4 Protein Resembles a Cell-Surface Receptor," *Nature* 325 (1987): 733–736.

4. R. Sherrington et al., "Cloning of a Gene Bearing Missense Mutations in Early-Onset Familial Alzheimer's Disease," *Nature* 375 (1995): 754–760.

5. E. Levy-Lahad, et al., "Candidate Gene for the Chromosome 1 Familial Alzheimer's Disease Locus, *Science* 269 (1995): 973–977.

6. D. O. Wirak et al., "Deposits of Amyloid Beta Protein in the Central Nervous System of Transgenic Mice," *Science* 253 (1991): 323–325.

7. J. Marx, "Major Setback for Alzheimer's Models," *Science* 255 (1992): 1200–1202.

CHAPTER 5

1. J. A. Hardy and G. A. Higgins, "Alzheimer's Disease: The Amyloid Cascade Hypothesis," *Science* 256 (1992): 184–185.

2. J. Hardy and D. J. Selkoe, "The Amyloid Hypothesis of Alzheimer's Disease: Progress and Problems on the Road to Therapeutics," *Science* 297 (2002): 353–356.

3. P. L. McGeer and E. G. McGeer, "The Inflammatory Response System of Brain: Implications for Therapy of Alzheimer and Other Neurodegenerative Diseases," *Brain Research. Brain Research Reviews* 21 (1995): 195–218.

4. M. Yegambaram et al., "Role of Environmental Contaminants in the Etiology of Alzheimer's Disease: A Review," *Current Alzheimer Research* 12 (2015): 116–146.

5. K. Maurer, S. Volk, and H. Gerbaldo, "Auguste D and Alzheimer's Disease," *Lancet* 349 (1997): 1546–1549.

6. W. J. Strittmatter et al., "Apolipoprotein E: High-Avidity Binding to Beta-Amyloid and Increased Frequency of Type 4 Allele in Late-Onset Familial Alzheimer Disease," *Proceedings of the National Academy of Sciences of the United States of America* 90 (1993): 1977–1981.

7. B. Wolozin et al., "Decreased Prevalence of Alzheimer Disease Associated with 3-Hydroxy-3-Methyglutaryl Coenzyme A Reductase Inhibitors," *Archives of Neurology* 57 (2000): 1439–1443.

8. J. Cummings et al., "Alzheimer's Disease Drug Development Pipeline: 2019," *Alzheimer's & Dementia* 5 (2019): 272–293.

9. G. Bartzokis, J. L. Cummings, D. Sultzer, V. W. Henderson, K. H. Nuechterlein, and J. Mintz, "White Matter Structural Integrity in Healthy Aging Adults and Patients with Alzheimer Disease: A Magnetic Resonance Imaging Study," *Archives of Neurology* 60 (2003): 393–398.

10. G. Bartzokis, "Alzheimer's Disease as Homeostatic Responses to Age-Related Myelin Breakdown," *Neurobiology of Aging* 32 (2011): 1341–1371.

11. G. Bartzokis, "Age-Related Myelin Breakdown: A Developmental Model of Cognitive Decline and Alzheimer's Disease," *Neurobiology of Aging* 25 (2004): 5–18.

12. S. Begley, "The Maddening Saga of How an Alzheimer's 'Cabal' Thwarted Progress toward a Cure for Decades," STAT, June 25, 2019, https://www.statnews.com /2019/06/25/alzheimers-cabal-thwarted-progress-toward-cure.

CHAPTER 6

1. C. R. Jack, Jr., et al., "NIA-AA Research Framework: Toward a Biological Definition of Alzheimer's Disease," *Alzheimer's & Dementia* 14 (2018): 535–562.

CHAPTER 7

1. J. Mervis, "Data Check: Federal Share of Basic Research Hits New Low," *Science* 355 (2017): 1005.

2. S. J. Nass, G. Madhavan, and N. R. Augustine, "Making Medicines Affordable: A National Imperative," in *Making Medicines Affordable: A National Imperative* (Washington, DC, 2017).

3. L. Fleming et al., "Government-Funded Research Increasingly Fuels Innovation," *Science* 364 (2019): 1139–1141.

4. C. Saez, "Study Shows Pharmaceutical Industry Investing in Basic Research, Some Questions Remain Open," Drug and Diagnostics Development, 2018; https://www .healthpolicy-watch.org/study-shows-pharmaceutical-industry-investing-in-basic -research-some-questions-remain-open.

5. E. Jung, A. Engelberg, and A. Kesselheim, "Do Large Pharma Companies Provide Drug Development Innovation? Our Analysis Says No," *First Opinion* (ed. STAT+) (2019).

6. M. Sullivan, Dr. Paul Aisen Q&A: Aducanumab for Alzheimer's, 2019, https:// www.mdedge.com/neurology/article/211030/alzheimers-cognition/dr-paul-aisen-qa -aducanumab-alzheimers.

CHAPTER 8

1. B. S. Ye et al., "Longitudinal Outcomes of Amyloid Positive versus Negative Amnestic Mild Cognitive Impairments: A Three-Year Longitudinal Study," *Scientific Reports* 8 (2018): 5557.

2. X. Chen et al., "Pittsburgh Compound B Retention and Progression of Cognitive Status—A Meta-Analysis," *European Journal of Neurology* 21 (2014): 1060–1067.

3. G. E. P. Box, W. G. Hunter, and J. S. Hunter, *Statistics for Experimenters: An Introduction to Design, Data Analysis, and Model Building* (New York: Wiley, 1978).

4. V. L. Villemagne et al., "High Striatal Amyloid β-Peptide Deposition across Different Autosomal Alzheimer Disease Mutation Types," *Archives of Neurology* 66 (2009): 1537–1544.

5. J. C. Dodart et al., "Immunization Reverses Memory Deficits without Reducing Brain Aβ Burden in Alzheimer's Disease Model," *Nature Neuroscience* 5 (2002): 452–457.

6. V. Volloch and S. Rits, "Results of Beta Secretase-Inhibitor Clinical Trials Support Amyloid Precursor Protein-Independent Generation of Beta Amyloid in Sporadic Alzheimer's Disease," *Medical Sciences* 6 (2018).

7. K. Herrup, "The Case for Rejecting the Amyloid Cascade Hypothesis," *Nature Neuroscience* 18 (2015): 794–799.

CHAPTER 9

1. R. Katzman, "Editorial: The Prevalence and Malignancy of Alzheimer Disease: A Major Killer," *Archives of Neurology* 33 (1976): 217–218.

2. B. E. Tomlinson, G. Blessed, and M. Roth, "Observations on the Brains of Demented Old People," *Journal of the Neurological Sciences* 11 (1970): 205–242.

3. B. E. Tomlinson, G. Blessed, and M. Roth, "Observations on the Brains of Non-demented Old People," *Journal of the Neurological Sciences* 7 (1968): 331–356.

4. Z. S. Khachaturian, "Diagnosis of Alzheimer's Disease," *Archives of Neurology* 42 (1985): 1097–1105.

5. G. McKhann et al., "Clinical Diagnosis of Alzheimer's Disease: Report of the NINCDS-ADRDA Work Group under the Auspices of Department of Health and Human Services Task Force on Alzheimer's Disease," *Neurology* 34 (1984): 939–944.

6. Khachaturian, "Diagnosis of Alzheimer's Disease."

7. A. Gomez and I. Ferrer, "Involvement of the Cerebral Cortex in Parkinson Disease Linked with G2019S LRRK2 Mutation without Cognitive Impairment," *Acta Neuropathologica* 120 (2010): 155–167.

8. B. T. Hyman and J. Q. Trojanowski, "Consensus Recommendations for the Postmortem Diagnosis of Alzheimer Disease from the National Institute on Aging and the Reagan Institute Working Group on Diagnostic Criteria for the Neuropathological Assessment of Alzheimer Disease," *Journal of Neuropathology and Experimental Neurology* 56 (1997): 1095–1097.

9. B. Vellas et al., "Long-term Follow-up of Patients Immunized with AN1792: Reduced Functional Decline in Antibody Responders," *Current Alzheimer Research* 6 (2009): 144–151.

10. C. R. Jack, Jr., et al., "Introduction to the Recommendations from the National Institute on Aging-Alzheimer's Association Workgroups on Diagnostic Guidelines for Alzheimer's Disease," *Alzheimer's & Dementia* 7 (2011): 257–262.

11. K. Herrup, "Commentary on 'Recommendations from the National Institute on Aging-Alzheimer's Association Workgroups on Diagnostic Guidelines for Alzheimer's Disease.' Addressing the Challenge of Alzheimer's Disease in the 21st Century," *Alzheimer's & Dementia* 7 (2011): 335–337.

12. Jack et al., "Introduction to the Recommendations."

13. B. T. Hyman et al., "National Institute on Aging-Alzheimer's Association Guidelines for the Neuropathologic Assessment of Alzheimer's Disease," *Alzheimer's & Dementia* 8 (2012): 1–13.

14. M. S. Albert et al., "The Diagnosis of Mild Cognitive Impairment due to Alzheimer's Disease: Recommendations from the National Institute on Aging-Alzheimer's Association Workgroups on Diagnostic Guidelines for Alzheimer's Disease," *Alzheimer's & Dementia* 7 (2011): 270–279.

15. R. A. Sperling et al., "Toward Defining the Preclinical Stages of Alzheimer's Disease: Recommendations from the National Institute on Aging-Alzheimer's Association Workgroups on Diagnostic Guidelines for Alzheimer's disease," *Alzheimer's & Dementia* 7 (2011): 280–292.

16. C. R. Jack, Jr., et al., "NIA-AA Research Framework: Toward a Biological Definition of Alzheimer's Disease," *Alzheimer's & Dementia* 14 (2018): 535–562.

17. G. P. Morris, I. A. Clark, and B. Vissel, "Questions Concerning the Role of Amyloid-β in the Definition, Aetiology and Diagnosis of Alzheimer's Disease," *Acta Neuropathologica* 136 (2018): 663–689.

18. M. D. Garrett, "A Critique of the 2018 National Institute on Aging's Research Framework: Toward a Biological Definition of Alzheimer's Disease," *Journal of Current Neurobiology* 9 (2018): 49–58.

CHAPTER 10

1. D. B. Friedman and T. E. Johnson, "A Mutation in the *age-1* Gene in *Caenorhabditis elegans* Lengthens Life and Reduces Hermaphrodite Fertility," *Genetics* 118 (1988): 75–86.

2. X. Song, F. Ma, and K. Herrup, "Accumulation of Cytoplasmic DNA Due to ATM Deficiency Activates the Microglial Viral Response System with Neurotoxic Consequences," *Journal of Neuroscience* 39 (2019): 6378–6394.

3. L. Hayflick and P. S. Moorhead, "The Serial Cultivation of Human Diploid Cell Strains," *Experimental Cell Research* 25 (1961): 585–621.

4. P. Sousa-Victor et al., "Geriatric Muscle Stem Cells Switch Reversible Quiescence into Senescence," *Nature* 506 (2014): 316–321.

5. H. M. Chow et al., "Age-Related Hyperinsulinemia Leads to Insulin Resistance in Neurons and Cell-Cycle-Induced Senescence," *Nature Neuroscience* 22 (2019): 1806–1819.

CHAPTER 11

1. M. D. Garrett, "A Critique of the 2018 National Institute on Aging's Research Framework: Toward a Biological Definition of Alzheimer's Disease," *Journal of Current Neurobiology* 9 (2018): 49–58.

2. G. M. McKhann et al., "The Diagnosis of Dementia due to Alzheimer's Disease: Recommendations from the National Institute on Aging-Alzheimer's Association Workgroups on Diagnostic Guidelines for Alzheimer's Disease," *Alzheimer's & Dementia* 7 (2011): 263–269.

3. R. Katzman, "Editorial: The Prevalence and Malignancy of Alzheimer Disease: A Major Killer," *Archives of Neurology* 33 (1976): 217–218.

4. McKhann et al., "The Diagnosis of Dementia due to Alzheimer's Disease."

CHAPTER 12

1. H. W. Querfurth and F. M. LaFerla, "Alzheimer's Disease," *The New England Journal of Medicine* 362 (2010): 329–344.

2. Y. Zhang et al., "An RNA-Sequencing Transcriptome and Splicing Database of Glia, Neurons, and Vascular Cells of the Cerebral Cortex," *The Journal of Neuroscience* 34 (2014): 11929–11947.

CHAPTER 13

1. S. Begley, "The Maddening Saga of How an Alzheimer's 'Cabal' Thwarted Progress toward a Cure for Decades," STAT, June 25, 2019, https://www.statnews.com/2019 /06/25/alzheimers-cabal-thwarted-progress-toward-cure.

2. C. R. Jack, Jr., et al., "NIA-AA Research Framework: Toward a Biological Definition of Alzheimer's Disease," *Alzheimer's & Dementia* 14 (2018): 535–562.

3. CalicoLabs, "We're Tackling Aging, One of Life's Greatest Mysteries," 2020, https:// www.calicolabs.com.

INDEX

Aβ peptide, 61, 65, 74, 80, 216, 219
Adacanumab, 128, 129
Aging, 169, 185
  evolution, 173–175
  metabolism, , 177–178
  oxidation, 176, 181
  research, 207–210
  senescence, 183–184
Alpha secretase, 63–65
Alzheimer's Association, 113, 154,
    164–166, 239–240
Alzheimer's disease
  2011 guidelines, 158–165, 188, 228
  2018 guidelines, 164–166, 188, 216, 228
  familial, 34–35
  mouse models, 66–69, 135–143
  preclinical, 162–165, 189, 229
  spectrum, 224
  sporadic, 35
  vaccine, 68–70, 139–145
Alzheimer's disease, definition, 3, 8,
    148–149, 188–189, 224–225, 228–229
  CERAD, 156, 188
  gold standard, 158–60, 228–229
  inflation, 10, 107, 150, 187–188

Amyloid, 25–28, 50, 55–56, 189–190,
    216, 218
Amyloid burden, 162
Amyloid cascade hypothesis, 73–96,
    131–148, 187, 199, 205, 211, 215,
    221, 228
Amyloid precursor protein, 61–65,
    74–83, 138, 216, 218
Anosognosia, 10
Antagonistic pleiotrophy, 174–175
APOE, 5, 37, 44–45, 85–87, 190
Astrocytes, 192–204, 218
Autism spectrum, 224

Bartzokis, George, 87–91
Bayer, Friedrich, 117
Beta secretase, 62–65, 90, 145 146
Beyruther, Konrad, 60
Biogen, 128–130
Biomarkers, 160, 220, 224
Blood pressure, 20–21, 200, 222
Blood vessel cells, 45, 55–56, 76,
    192–204
Braak, Heiko, 153, 163
Butler, Robert, 105

Calcium, 47–49, 75–76, 80, 82
Central Dogma of Alzheimer's disease,
  71–72
Central Dogma of molecular biology,
  56–59, 63, 179
Central Dogma of the pharmaceutical
  industry, 118
Cholesterol, 85–87, 220
Cholinergic hypothesis, 50–51
Clinical trials, 119, 126, 187, 235
  prevention, 42
  therapeutic, 42–43

Dementia, 10, 110, 189, 213, 221, 224
  frontotemporal, 224
  Lewy body, 190, 224
  progressive supranuclear palsy, 190
  vascular, 10, 190, 199–200, 223
Diabetes, 19–20, 184, 188, 220, 221–222
Disease gene, 37
DNA damage, 179–180, 208–210
Down syndrome, 55–56, 59–60, 77, 82
Drug development, 119–130, 215, 235

Education, 18–19
Executive function, 20, 189

FINGER, 20

Gamma secretase, 62–65, 90, 145–146
Garrett, Mario, 166, 188
Glenner, George, 54–56, 59

Hardy, John, 73–79, 92
Hayflick, Leonard, 183
Higgins, Gerald, 73–78, 92
Hippocampus, 199

Infection, 49–50
Inflammation, 17–18, 40, 84, 180–181,
  193, 213, 218
Insulin, 19–20, 38–39, 57, 61, 177–178,
  184, 210, 221–222, 240–241

Journals, 240–241

Katzman, Robert, 105, 150, 157, 188
Khachaturian, Zaven, 105, 155, 157
Kinyoun, Joseph, 103
Kraepelin, Emil, 25–27, 31, 51, 63–65,
  73, 76–77, 106, 149, 188

Lipids, 43–45, 85–87, 90
Lysosomes, 46, 75–76, 80, 82, 91–92

Medicare, 14
Mediterranean diet, 19, 39
Microbiome, 222
Microglia, 41, 192–204, 213–214,
  217
Mild cognitive impairment, 133,
  161
Mitochondria, 46–47, 92, 112, 164,
  178–179, 198, 230, 240
Müller-Hill, Benno, 60
Multiple sclerosis, 104, 214
Myelin, 45, 48–49, 87–91, 204–205,
  211–215, 219

National Institute on Aging (NIA), 149,
  154, 158, 164–166
  funding, 229–233
  history, 105–114
  research, 107–114
National Institute of Neurological
  Diseases and Stroke (NINDS),
  104–105 108, 112, 230
National Institutes of Health, 101–113,
  121, 149, 215, 227
  research funding, 99–114, 121
Neighborhood model, 192–204, 209
  aging, 198–204
  burn-in, 199–204, 209, 220–223
  pixel, 196, 198, 204
Nerve cell death, 76
NSAIDs, 17–18, 40–43, 222
Nun's study, 18

Oligodendrocytes, 192–205, 212–215, 218

Oxidation, 80, 92–94, 176

Parkinson's disease, 10, 94, 104–105, 112, 153, 156, 199

Perry, George, 93

Pharmaceutical industry (PhRMA), 115–130, 166, 215
  government partnership, 227–239
  history, 117–118
  research, 121–130, 234–237
  university partnership, 233–236

Pittsburgh compound B, 133, 140

Plaques, 25–28, 50–54, 62–65, 68–69, 84, 107, 113, 151–158, 189, 220

Preclinical Alzheimer's, 229

Presenile dementia, 11, 31–32, 107, 149–150, 154–155

Presenilin, 64–65, 81, 90–91, 136, 138

Press, 241

Risk factor gene, 37

Roche Institute of Molecular Biology, 123

Schenk, Dale, 68–69

Selkoe, Dennis, 79, 92

Senile dementia, 10, 150–154

Smith, Mark, 93

SNP (single nucleotide polymorphism), 35, 38

Statins, 18, 86–87, 110

Synapse, 50–51, 86, 138, 172, 191

Tangles, 75, 95, 107, 220

Tau, 75–76, 80, 84, 90–91, 95, 113–114, 160–161, 169, 220, 232–233, 241

Tomlinson, Bernard Evans, Gary Blessed, and Martin Roth, 150–154

University research, 234, 236

Vesicles, 44, 46–47, 76, 90–92

Wong, Caine, 54–56, 59